Corvette C5
Performance Projects
1997–2004

Corvette C5
Performance Projects
1997–2004

Richard Newton

MOTORBOOKS

First published in 2005 by Motorbooks, an imprint of MBI Publishing Company, Galtier Plaza, Suite 200, 380 Jackson Street, St. Paul, MN 55101-3885 USA

© Richard Newton, 2005

All rights reserved. With the exception of quoting brief passages for the purposes of review, no part of this publication may be reproduced without prior written permission from the Publisher.

The information in this book is true and complete to the best of our knowledge. All recommendations are made without any guarantee on the part of the author or Publisher, who also disclaim any liability incurred in connection with the use of this data or specific details.

This publication has not been prepared, approved, or licensed by General Motors. We recognize, further, that some words, model names, and designations mentioned herein are the property of the trademark holder. We use them for identification purposes only. This is not an official publication.

Motorbooks titles are also available at discounts in bulk quantity for industrial or sales-promotional use. For details write to Special Sales Manager at MBI Publishing Company, Galtier Plaza, Suite 200, 380 Jackson Street, St. Paul, MN 55101-3885 USA.

ISBN-13: 978-0-7603-2081-5
ISBN-10: 0-7603-2081-0

Editor: Lindsay Hitch
Designer: Christopher Fayers

Printed in China

On the front cover: The C5 surprised a lot of people who were accustomed to the rough-riding Corvettes of the past.

On the title page: The C5 Corvette looks good from any angle.

On the back cover: All of the aftermarket induction systems pick up air in the same place—in front of the radiator.

About the author: Corvette expert Richard Newton is an ASE Master Technician who spent 15 years in the auto service industry before becoming a magazine editor and freelance automotive writer. Newton is the author of *101 Projects For Your Corvette 1984-1996*, *How To Restore & Modify Your Corvette 1968-82*, *Corvette Restoration Guide 1963-1967*, and *Ultimate Garage Handbook*. He resides in Bonita Springs, Florida.

Contents

Introduction 7
Tools of the Trade 10
Deciphering the Information Boxes 16

Section One: Basics 18
   Project 1: Raising Your C5 18
   Project 2: Changing the Engine Oil 21
   Project 3: Replacing the Serpentine Belts 24
   Project 4: Replacing the Air Filter 27
   Project 5: Replacing the Fuel Filter 29
   Project 6: Detailing Your Corvette 31

Section Two: Engine 34
   Project 7: Cleaning the Throttle Body 34
   Project 8: Replacing the Thermostat 38
   Project 9: Reprogramming Your Corvette's Computer PROM ... 40
   Project 10: Replacing the O_2 Sensors 42
   Project 11: Changing the Coolant 44
   Project 12: Dressing up the Engine Compartment 46
   Project 13: Replacing the Spark Plugs and Spark Plug Wires ... 49
   Project 14: Installing an Alternative Intake Manifold 52
   Project 15: Replacing the Fuel Injector 55
   Project 16: Installing a Louder Exhaust System 58
   Project 17: Installing Headers 62

Section Three: Electrical 66
   Project 18: Replacing Your Starter 66
   Project 19: Replacing the Alternator 67
   Project 20: Upgrading Your Audio System 69

Section Four: Drivetrain 73
   Project 21: Installing a Short-Throw Shifter 73
   Project 22: Servicing the Six-Speed Manual Transmission .. 75
   Project 23: Servicing the Automatic Transmission (Fluid Change) ... 78
   Project 24: Selecting a Shift Kit for the Automatic 83
   Project 25: Installing a Skip Shift Eliminator 85
   Project 26: Changing the Rear Differential Lubricant 86

Section Five: Brakes 88
Project 27: Bleeding and Flushing the Hydraulic System 88
Project 28: Replacing the Brake Pads and Rotors 90
Project 29: Installing Stainless-Steel Brake Hoses 92
Project 30: Changing the Brake Bias 94
Project 31: Replacing an ABS Sensor 96

Section Six: Suspension 97
Project 32: Lowering Your C5 97
Project 33: Installing Polyurethane Bushings 99
Project 34: Replacing the Shock Absorbers 101
Project 35: Choosing a Sway Bar 103
Project 36: Selecting Tires and Wheels 105
Project 37: Performing an Alignment 108
Project 38: Replacing the Sway Bar End Links 111

Section Seven: Cosmetics 113
Project 39: Customizing the Interior Trim 113
Project 40: Installing a Body Package 117

Appendix 120
Index 127

Courtesy of General Motors

Introduction

The C5 has become the Corvette of choice for high-speed fun. It's easily modified and there are a lot of parts available. As the price of used C5s depreciates, look for even more C5 race cars and dedicated track events, not to mention Solo 2 cars.

They called it the C5 Corvette, and it put the excitement back into the Corvette hobby. It was as if the Corvette had begun all over again. The C4 (1984–1996) had been around forever. General Motors had every intention of ending the C4 much sooner than it did. But when GM ran into some rather serious financial difficulties, everything—especially the Corvette—was placed on hold.

This was the first time that GM rolled out a new engine, a new frame, and a new body design all at the same time. Traditionally, GM revised only parts of the Corvette when it introduced a new generation.

In 1953, we got the basic six-cylinder passenger-car engine with the traditional passenger-car frame. The only really new part was the fiberglass body. Even the legendary 1963 Corvette had the same engine lineup as the 1962 Corvette. This was repeated when the '68 Corvette was introduced with the engine lineup from the '67 Corvette. Also keep in mind that the Corvette used the same basic frame and suspension from 1963 to 1982. Only the body and the interior were changed, in 1968.

In 1997, we got the whole deal all at once. There was one downside, though. This new Corvette was a few years late, as GM had flirted with bankruptcy during the C5's development process. That meant that even though we finally got an all-new Corvette, it was a little out of date even as it was introduced.

The biggest change was that the Corvette was leaving its hot rod heritage in the past. All the rough edges were removed, and the Corvette became a high-speed grand touring car. The days of raw power and little foibles, or "character," as some people said, were left behind. The benchmark for the C5 Corvette was Lexus. GM had turned the Corvette into a very fast Lexus. Sales were phenomenal. The C5 set new records every year it was produced. The Corvette did so well that even the corporate bean counters were happy.

The C5 shocked a lot of people who were into the rough-riding Corvettes of the past. It didn't make enough noise for a lot of traditional Corvette fans. Even the styling was controversial. Nonetheless, sales continued to exceed expectations for the entire production run.

GM had rejected earlier C5 designs that it had determined were too advanced, thinking the public was not ready for them. Essentially, the C5 Corvette was focus-grouped to death. It was designed in an era when the marketing people

This Corvette was created by Pratt and Miller as a way of getting the C5 homologated by the Le Mans officials. It was brought to Sebring for examination and hasn't been seen since.

Some people like side spears and some don't. Ironically, after they were introduced, GM discovered that they actually aid the laminar side airflow.

were deathly afraid of upsetting a potential customer. In their effort to please everyone, they ended up with a car that was a little too bland for the traditional Corvette owner.

Then again, GM wanted to move past the traditional Corvette owner and bring new customers into the dealerships. There simply weren't enough traditional Corvette customers to justify the expense of developing an all-new Corvette. Fortunately for GM, this new Corvette concept worked.

The convertible followed close on the heels of the coupe. Never before had GM designed the coupe and convertible at the same time. For the first time, Corvette customers got a really solid convertible that didn't flex and twist as it went down the road.

Then came the fixed-roof coupe (FRC), which was also designed at the same time as the coupe. But problems arose as the mission for the FRC continually changed. The car was originally designed as the "cheap Corvette." That idea was killed early on, and the FRC suddenly became the performance version. GM decided that this model, and this model only, would have the optional Z06 performance package.

Corvette performance returned with the Z06—truly a different animal than the base Corvette. All of the complaints about the Corvette being a Lexus—or worse yet, a Cadillac—disappeared with the introduction of the Z06 option.

Amazingly, it took GM a long time to bring the Z06 option to market. The Z06 had been developed as part of the original C5 package. I had the chance to drive one in 1997 and I loved it. If GM had made this package available earlier, it would have silenced all the critics. Despite the Z06 delay, Corvette sales just kept moving along.

To choose between the three different Corvettes in the C5 generation, you have to make a decision about how you

A big bass speaker and a bottle of nitrous—it doesn't get much wilder than this.

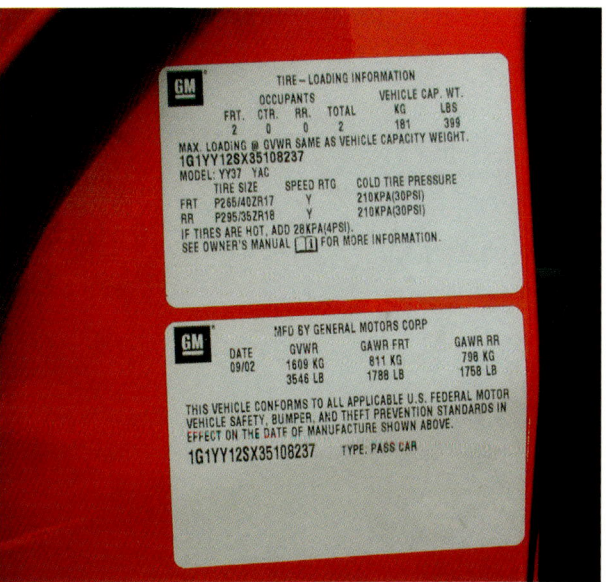

There's a wealth of information on the door pillar. If your Corvette doesn't have these labels, it may have been in an accident.

intend to use your car. A lot of you already own a 1997–2004 Corvette, so that decision's already been made. If you haven't purchased a car yet, you have to think about what you really want out of your Corvette.

The C5 convertible is a wonderful cruising car for the top-down enthusiast. The best part is that the convertible is a really solid car. Earlier Corvette convertibles were almost afterthoughts for GM. The only way GM could hold the C4 convertible together was to bolt a huge X-brace to the bottom of the car, and it still flexed like a wet piece of lasagna.

But the C5 put all that in the past. When Dave McLellan and his merry band created the C5 convertible, they knew they needed a solid car. The convertible was actually designed first, making the C5 both a convertible and a high-performance street car.

The best part is the C5 convertible has a real trunk, so you can use it for long trips. Noise is also not an issue; when the top is up, the car is almost as quiet as the other two models. This convertible is really a winning option.

The traditional, base-model Corvette has a huge glass panel that covers the rear compartment. This Corvette is the most common and the one that can benefit the most from basic modifications.

This base-model C5 features a removable roof panel that makes the car almost a convertible. One person can remove the top easily; it's not a structural part of the car, and it's attached with simple latches and not bolted in like the C4. If you just want a nice Corvette, the hardtop with the removable roof panel is probably your best choice. It's faster than anything you really need, and the car is dead reliable.

If you're serious about having a fast Corvette, set your sights on a fixed-roof coupe with the Z06 package. Since the Z06 engine was offered with only the coupe, it's a simple decision. This is the Corvette for the person who intends to spend time on the racetrack, be it a drag strip or road course.

Working on your Corvette is a major part of the Corvette experience. It's nice to drive your Corvette, and it's fun to have people admire it, but working on it is one reason you bought the car in the first place. "Working" on it can mean everything from waxing the car to installing a new 427-ci LS6 engine. Owning a Corvette is a very personal experience. That experience becomes even more personal as you work on your Corvette. Just as the mother duck bonds with her ducklings, you bond with your Corvette.

The C5-R Corvette was the very first Corvette to win a major race when it captured overall first place at Daytona in 2001. The C5 has been the most successful car in the history of Corvette racing.

Tools of the Trade

Most of us are tool junkies. It seems that anyone who likes cars has an equally serious addiction to tools. That's why, when the family goes shopping at the local mall, most of us find a way to sneak off to the Sears tool department. You can never own enough tools.

By the time you get around to owning a Corvette, you'll probably already have amassed a huge number of tools. No matter how you came to acquire them, you likely haven't thrown out a tool since the 10th grade. You may have lost a number of tools, but you've never willingly disposed of a single wrench or screwdriver.

Virtually every hand tool sold today comes with a lifetime warranty. The only concern is how easy it is to use that warranty. When I was in the professional end of the business, I hated the Sears warranty and loved the Snap-on warranty. If I broke a tool in my shop, the Snap-on dealer would replace the tool on his next weekly visit. Now that I'm no longer in the service business, the Snap-on truck no longer arrives on a weekly basis and it's easier to replace the Craftsman. Nothing is as good as a Snap-on tool, but nothing is as expensive as a Snap-on tool. No weekend warrior actually needs Snap-on tools, but they can really make you feel good while you work on a car.

Screwdrivers: I own more than 50 screwdrivers and regularly use the Snap-on or Matco models. Screwdrivers are not a place to save money. Buy screwdrivers with hardened tips and you'll get a lifetime of use out of them. Even the handles feel better. You'll seldom strip a Phillips-head screw with a quality screwdriver.

Electric screwdrivers should not be used to break screws loose, but they can make fast work of interior screws. For this purpose, the cheaper store-brand versions work just fine.

Adjustable Wrench: I can't remember using an adjustable wrench on a Corvette. If you buy one, make sure it's something you can use around the house since you won't need it for your Corvette.

Pliers: Much like screwdrivers, you can never really have enough pliers. The big difference is that good pliers are a lot more expensive than good screwdrivers. The most useful pliers for Corvette work are long needle-nose pliers.

Sockets and Drivers: You can spend more money on sockets and drivers than you paid for your Corvette. We start with the three basic sizes for drivers: ¼ inch, ⅜ inch, and ½ inch. Then there are shallow sockets and deep sockets, and a new

Make sure that all of your screwdrivers have hardened tips, and invest in tools that will last a lifetime. With high-end screwdrivers, you can replace just the blade portion of the tool.

You can never own enough pliers, and they're available in a wide variety.

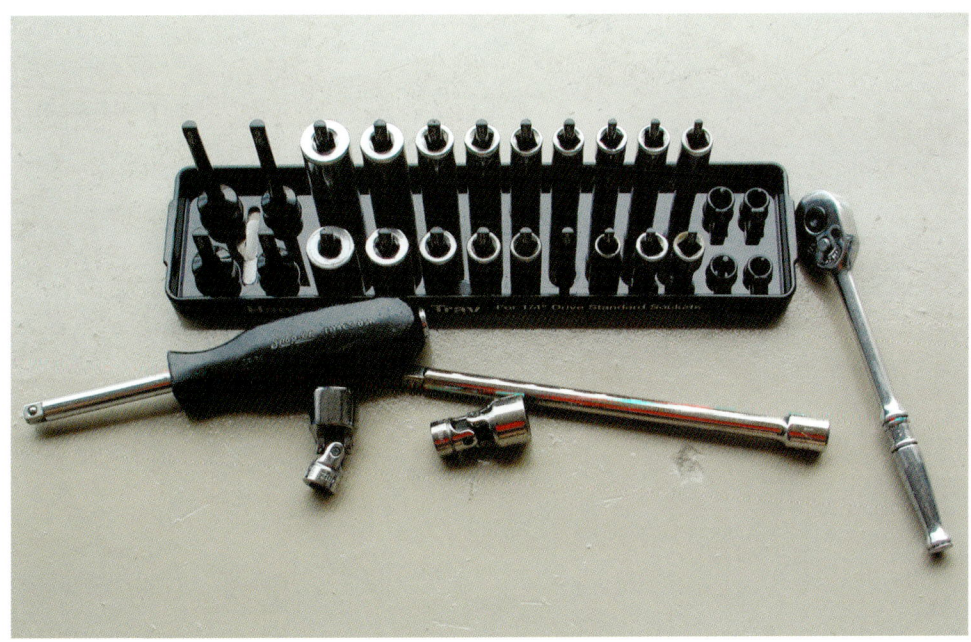

Sockets are fun. This is a decent-size set of ¼-inch drive deep sockets. Having tried every version of socket organizer in the world, this type of tray, available from Sears, is my favorite. It's easy to bring over to the car, and it's easy to keep clean.

size that can best be described as "in between." Every single socket comes in a 12-point version or a 6-point version. The good news is that no one ever purchases them all at one time. The bad news is that you need all three sizes of sockets to work on the C5 Corvette.

I'm a huge fan of ¼-inch drive sockets, but you can get by with ⅜-inch drive sockets for most of your work. Craftsman makes nice sockets at reasonable prices. I suggest that you start with the basic ⅜-inch drive metric set and add a few sockets on either end of the range for complete coverage.

Remember that a deep socket will always work where you could use a short socket, but if you need a deep socket then that's the only thing that will suffice. Strive for a good collection of deep 12-point sockets with a couple of ⅜-inch drive ratchets.

You should also consider getting several socket drivers. You'll want the basic short one and a longer one with a swivel head. A long breaker bar is also nice to have for those stubborn bolts.

Hammers: The two most important hammers are the soft face and 10-pound mini sledgehammer. My rawhide mallet is probably the most-used hammer in my collection. I prefer the rawhide mallet to plastic because it leaves fewer marks, but it has no muscle. The 10-pound mini sledgehammer can provide all the muscle you'll ever need on a Corvette. You can buy a basic steel version or the plastic-covered version. They're great for taking apart the rear suspension on the C5. I also love to use a 10-pound hammer as a rotor-removal tool.

Timing Light: You will not need a timing light for work on a late-model Corvette. Instead, buy more ¼-inch metric sockets.

Torque Wrenches: Buy a really good torque wrench and have it calibrated every couple years. Don't allow this precision instrument to bang around in your toolbox. You'll use your torque wrench to check the lug nuts when you remove and replace your tires.

Warped brake rotors are often the result of not tightening the wheels to the proper torque setting. A good ½-inch-drive

Buy more hammers. Big hammers are great for knocking rotors off the hub of your Corvette. The rawhide ones are great, because they don't leave marks.

11

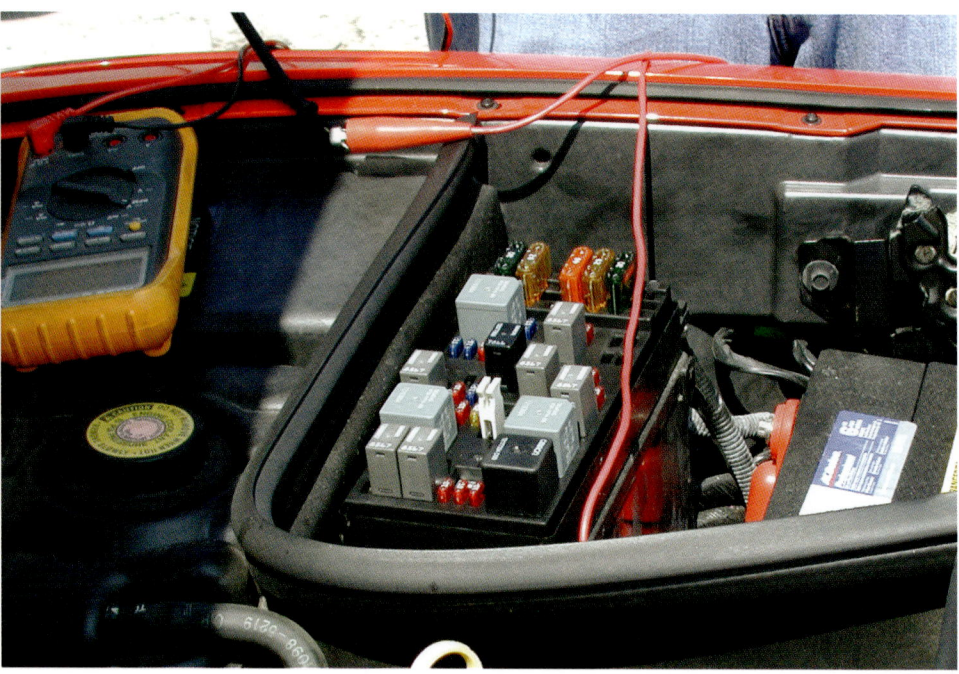

A volt ohm meter, often called a "VOM," is very handy to have around a Corvette. A basic model, which is usually under $50, works well for the average hobbyist.

torque wrench costs nearly the same as one new Corvette rotor. Spend the money on a good torque wrench rather than brake rotors.

Most of the places where you need a torque wrench on the Corvette require a ½-inch drive torque wrench. But you might add the ⅜-inch drive torque wrench to your Christmas list.

Wrenches: If you don't own any wrenches at all, purchase a set of metric combination wrenches. These wrenches have a box on one end and an open end on the other. The box end is used for breaking nuts loose, or for tightening, and the open end is best for getting a nut all the way off the bolt once you've broken it loose. Beyond the basic combination set of wrenches, there are a lot of options. Remember, you can never have too many wrenches.

Line Wrenches: Line wrenches are simply a variation of the open-end wrench. They're used to loosen and tighten hydraulic or fuel lines. The opening is a lot smaller than that of a conventional open-end wrench. Hydraulic line fittings are usually very tight, which is why you want a line wrench. Line wrenches are also sometimes called "flare nut wrenches." You'll need three of them in metric sizes if you plan on replacing the master brake cylinder or the brake caliper lines.

I've used every line wrench on the market and recommend that you purchase a Snap-on set. The jaws of the wrench won't spread apart, which could cause the corners of the fitting to round off. The jaws of the less-expensive wrenches actually spread, defeating the whole purpose of a line wrench.

Lighting: The C5 Corvette has more nooks and crannies than any car I've ever owned. A really good flashlight lets you shine a finely focused light on the areas under examination.

If you don't already own one, a fluorescent extension cord light will come in handy. I prefer the short version, as it's more likely to fit into the space where you need the light. The longer types are nice if you're doing any engine work where a lot space in available. They can just hang from a hood latch and light the entire engine bay.

Portable halogen lights throw out a tremendous amount of light. They're also very cheap at home improvement stores. They actually get hot enough, however, that it can be uncomfortable working around them. This limits their usefulness. They're best for use under the car when you're doing suspension or brake work.

Electrical Repair: The electronics in the C5 are complicated enough that you really need a lot of skill before you plunge into the electrical system. The factory manual has an outstanding section on electrical repairs that you should read at least twice before you mess around with any of the wires on your Corvette.

The most important tool is the multimeter, which measures voltage, resistance, and amps. You can purchase a good one for under $50. Learn how to use it, and you'll be surprised how many problems you can solve.

Getting the insulation off a wire is a real trick without a wire stripper. A top-quality wire stripper is quite inexpensive and will last a lifetime.

Diagnostic equipment is now cheap enough that you might consider purchasing a used system for your car. But these

If you own a Corvette, you're going to need Torx sockets. This whole set was cheaper than three separate Torx sockets would have cost.

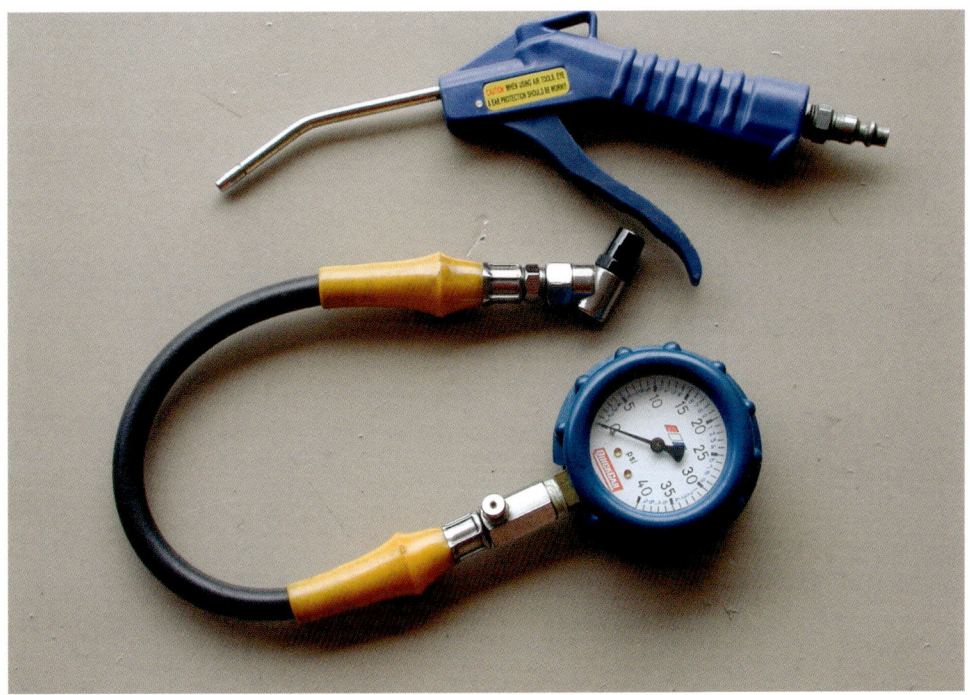

Here are two handy air tools. The long-nose air gun is especially nice for detailing engines and blowing the dust out of the dash vents. The tire gauge is a very serious bleed-down model with an easy-to-read large dial. You can just fill your tires and bleed the air down until you reach the appropriate pressure.

tools require a lot of skill to use effectively—even the largest dealerships only have one or two people skilled at using diagnostic equipment.

Battery Charger: A battery charger can prevent you from having to get the wrenches out to replace the battery and alternator every few years. Corvette batteries need to remain fully charged during those long winter months when you don't drive the car. I prefer the Battery Tender device to a battery charger for this task, but you can have both in your garage. They both make sense and they're both inexpensive.

PDA and Diagnostics: There was a time when you needed a huge diagnostic console to know what was going on with the

13

You'll need a dial gauge to check the brake rotors. As often as the C5 seems to warp rotors, this tool is going to be well used. Purchase a gauge with multiple mounting points that is accurate within 0.001 inch.

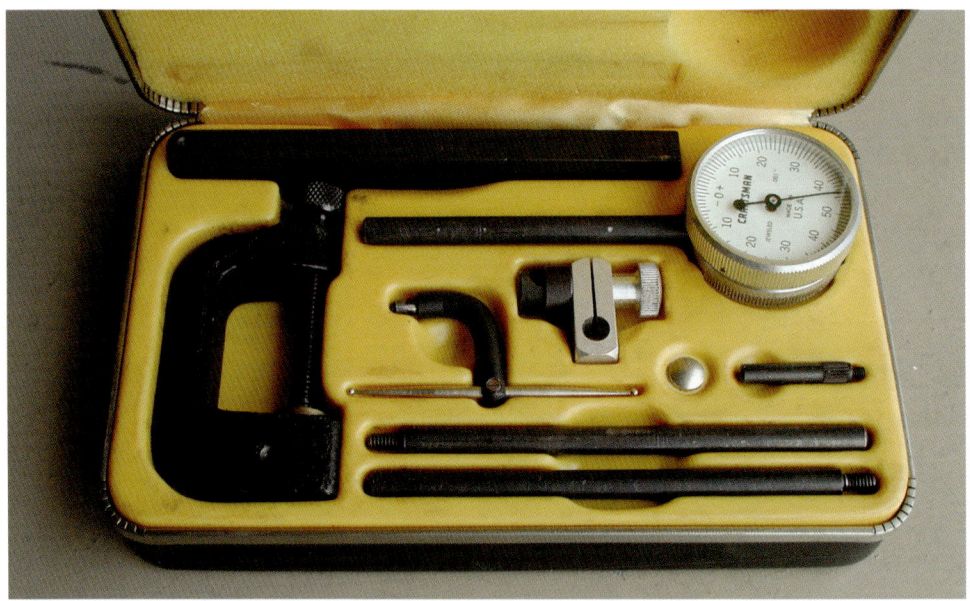

You won't need one of these diagnostic tools for your home garage, but they're fun once you learn how to use them. You can find this professional-grade tool on eBay for around $2,000. Your local Corvette shop should have a GM Tech 2, and more importantly they should know how to use it.

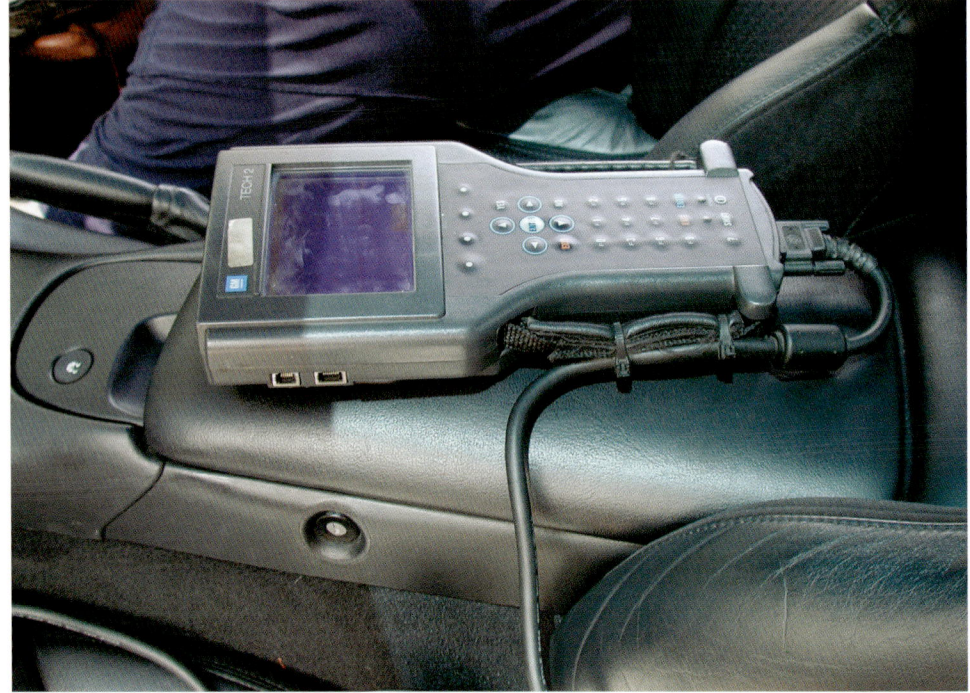

engine in your Corvette. Now you can check everything out with a hand-held personal digital assistant (PDA). You can purchase diagnostic software for your PDA and run a series of basic tests. Combined with a personal computer, you'll have an affordable, flexible, and upgradeable automotive diagnostic system.

Air Compressor: I can't imagine working in a garage without an air compressor. You don't need a gigantic unit, just one that allows you to fill tires and run a few air tools. If you have a bead blasting cabinet in your garage, though, you'll need a fairly large compressor to produce the required volume of air. When you purchase an air compressor, think of future ways you might use the compressor. Get a compressor that can be plugged into a normal 120-volt home electrical supply. A vertical model will take up a lot less space on the floor of your garage. You'll need about 25 feet of air hose, a good-quality blowgun nozzle, and an air chuck for filling your tires.

Lights are a big deal to me. The light on the left is battery-operated, which means I don't have extension cords all over the garage floor, and the same battery can also be used in my drill. The great thing about modular sets is that if you purchase wisely you can use one battery for a variety of different tools.

You need to be concerned about two items when you purchase a compressor. Every air tool has a cfm (cubic feet per minute) rating. A ½-inch impact wrench only uses about 4 cfm at 90 psi. On the other hand, the average bead blasting cabinet uses 15–20 cfm at about 80 psi. If your compressor can't supply enough air, you'll have to stop every few minutes to let the air build up again.

Pay attention to the psi (pounds per square inch) rating of the air compressor. This simply measures how much air you can compress into your air storage tank. The tighter you can pack the air into the tank, the more air you'll have available for use.

Remember, cfm is about power, and psi is about storage. If you don't think you'll ever own a bead blasting cabinet, you can find some really inexpensive compressors. On the other hand, you'll need to spend *at least* $500 for a compressor good enough to run a blasting cabinet. Most air compressors will last over 20 years in a home garage, so plan for the future.

The Advanced Collection

This is where you can spend some serious money for tools—tools that are critical but seldom used.

Dial Indicators: Dial indicators are most useful for brake work. You can't check brake rotor runout (warpage) without a good-quality dial indicator. When you shop around, keep in mind that the mounting system is more important than the actual dial gauge. When you check your brake rotors, you want the dial positioned at 90 degrees to the face of the rotor. Make sure you purchase a mounting system that will allow you to do this.

The real question is whether to purchase a dial indicator that reads out in metrics or one that reads out in the American system. The C5 Corvette is a totally metric car, but the factory manual gives specifications in both millimeters and inches. Base your decision on what other cars you have in your garage. If your other toy is an old straight-axle Corvette, get one that reads out in inches. If there's a Porsche in your garage, get the metric version.

Micrometers: The only reason to have a micrometer around is to measure brake rotor thickness. There's a special micrometer for this task, but a normal micrometer will do just fine.

The biggest problem with micrometers is that they usually have a range of only 1 inch, meaning you'll often need several sizes. Since the C5 Corvette front brake rotor is 1.205 inches thick and the rear is 1.02 inches thick, you'll be able to get by with a micrometer that reads from 1 to 2 inches. You might even consider one with a digital read out if you aren't familiar with micrometers.

Dial Caliper: Dial calipers are easy to use and can measure everything from a few thousandths of an inch to 6 inches. Dial indicators can measure anything from pieces of sheet metal to the diameter of an exhaust pipe. Once again, the digital readout models are a lot more convenient.

Engine Tools: Skip all of the traditional engine-building tools like ring compressors and valve spring compressors. You can purchase a crate motor for far less than it costs to rebuild your own engine.

Air Tools: Air tools exist for one reason—they make the job go faster. The big problem with air tools is that you need an air hose and a compressor. You don't need air tools to work on your Corvette, but if you want to use air tools, purchase a ⅜-inch air ratchet. Air ratchets are great for running nuts down on bolts and are small enough that they don't get in the way like an impact gun.

There's really no need for an impact gun in the home shop. Impact guns are expensive, and you can do a lot of damage if you aren't careful. If a bolt is really tight, you can usually remove it with a 24-inch breaker bar placed on your ½-inch socket.

Digital Camera: Strongly consider getting a digital camera before you start taking things apart. There's a good chance that your project will drag out several days or weeks beyond your original time schedule, and the pictures can help you put things back together. Just remember that it's better to have too many pictures than too few.

Deciphering the Information Boxes

At the beginning of each project you'll find a list of topics to help you decide if you really want to tackle the project or leave it for another time—or another person. The information boxes describe what's involved and what you'll need in the way of tools, knowledge, and money.

Time: "Time" is a rough estimate of how long the project would take a person with reasonable skills and knowledge. General Motors has a flat-rate manual that lists every possible task you can perform on a Corvette, and even that list isn't always right. A good service technician can beat the flat-rate book all day long. Another technician, working in the same dealership, will have trouble making a decent living because every job takes him two to three times longer than the time listed in the GM flat-rate manual.

I generally estimate that everything I do to my Corvette will take twice as long as I originally thought. At least that sort of thinking will get you in the ballpark.

Tools: "Tools" will give you a rough idea of which tools you'll need for each project. For the simple projects, I've listed the wrenches and sizes you might need. When it comes to the advanced projects, I've only listed the specialty tools. For engine removal you'll need a toolbox that's got a lot of sockets and wrenches. No matter what tools you already own, you'll make several trips to the tool department for more. You might want to add the cost of new tools into the budget for a given project.

Talent: "Talent" is an estimate of the skill level you will need to complete a project. The range of abilities for weekend-warrior mechanics is tremendous. Be honest with yourself before you take on a project. If a project has one star icon, anyone can give this project a shot. Two stars mean that you should be slightly proficient with wrenches. When we get to the third level, you need to have enough experience to do things with very little guidance. Changing the intake plenum on an LS1 engine might fall into this category. When you get to the fourth level, you had better know what you're doing and have a lot of tools in your toolbox. These are not projects for the masses. Generally, level four and five jobs are best left to the professional, or the very brave.

Applicable Years: Sometimes projects will apply only to certain model years. Fortunately, there aren't a lot of variations in the 1997–2004 Corvettes and most of the projects in this book will

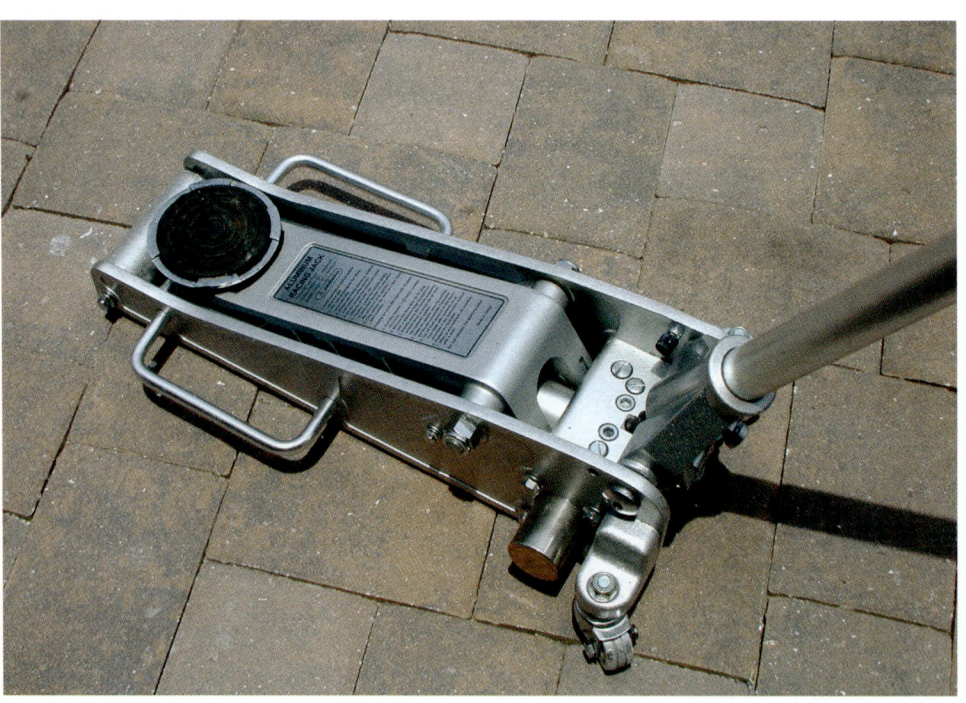

Ask around to find out what jacks people are using and which they're happy with. This type of jack is all the rage, but many people aren't happy with it. While it only takes a couple of strokes to raise the car, you have to use a tremendous amount of force.

If you're just getting started, there are some wonderful tool sets available. This one was purchased from Costco and the quality is amazing. And the price is even lower than at Sears. If you shop around, you can find some great bargain sets and some incredible junk—just be careful.

apply to all eight years. There were really only two engines—the LS1 and the LS6—and they're close enough that they might as well be the same. The chassis is the same for all the years the C5 was in production, which makes life a lot easier for everyone.

Tab: The tab is a ballpark idea of how much money you're going to spend. Over the years, I've developed a formula for project budgets. I look up all the prices for a project in the various catalogs and make a detailed list of the parts and prices. Then I take that total and double it. In my experience, every project takes on a life of its own and exceeds the preliminary budget at least twofold.

Torque: When your Corvette was assembled in Bowling Green, Kentucky, every single nut and bolt was set to a very specific torque setting. You don't have to be as precise as the factory, but don't disregard the proper torque settings for your Corvette. You need to be diligent with the specifications for the brake system, the chassis, and any engine work you might perform. I've only listed the most critical torque settings in this book. Refer to the factory manual for any other specifications.

Tip: I've included some little tips that will help you through a particular job. In most cases, it's something that a technician discovered years ago, and in some cases it may be something that many of you already know about.

Performance Gain: There's very little you can do to improve the performance of the C5 Corvette in one area without detracting from another area. I'll explain what gains you can expect from a modification and how it will affect another aspect of your Corvette.

Complementary Work: As soon as you replace one part on your Corvette you'll notice two other areas that might need some help. Lots of times, if you're working in one area you should consider working on an adjacent part. In combining projects, the real trick is knowing when to stop. I've included complementary projects where they make sense. There are some stand-alone jobs on your Corvette; enjoy these jobs for what they are. Don't try to stretch every task into a series of tasks. There's also a work/drive ratio. The longer you work on your Corvette without driving it, the quicker you'll lose interest in the car. Remember, you originally bought your Corvette to drive. When the fixing-up time exceeds the driving time, your interest will wane. Then your Corvette gets sold to a person who still has the passion that you once exhibited.

SECTION ONE
BASICS

Projects 1-6

PROJECT 1	Raising Your C5

Time: 10 minutes

Tools: Hydraulic floor jack, jack stands   **Torque:** N/A

Talent: ★   **Applicable Years:** 1997–2004   **Tab:** $75

Tip: AutoZone and Wal-Mart usually have the best prices and the largest selection for jacks and jack stands.

Performance Gain: N/A

Complementary Work: N/A

The C5 came from the factory without any visible jacking points. According to the shop manual, the proper points to lift the car with a four-point hoist are the holes that were used to attach the chains on the transporter. These holes are located on the frame rails of the car and in approximately the same locations as the jacking points on the C4. But the body of the C5 wraps down the side of the car and rolls under the frame rails. The body actually covers a great deal of the frame rails.

There are small cutouts in the plastic rocker panels for access to these openings, but they help only a little. Most floor jacks (also four-point lifts and jack stands) have pads larger than these openings, meaning that the lifting pad too often rests on part of the body, not the steel frame. This will crush the body panel between the frame and the lifting pad of the jack, hoist, or jack stand.

You can lift your C5 with simple garage tools in a few minutes. There are two different techniques—the "jacking puck" method and the wooden block method.

Lifting Your C5 with Wooden Blocks
To lift your C5 using wooden blocks, you'll need a small hydraulic floor jack; four small jack stands; two pieces of 2x6 wood, 25 inches long; and four pieces of 2x12 wood, 16 inches long.

Start by placing one of the 2x12x16-inch pieces of wood in front of each tire, both front and rear. Don't set the wood right against the tire, but rather place the boards a few inches

It's almost as if GM designed these front and rear crossmembers as jacking points. While GM recommends other points, it didn't take Corvette owners long to figure out that the crossmembers work really well.

18

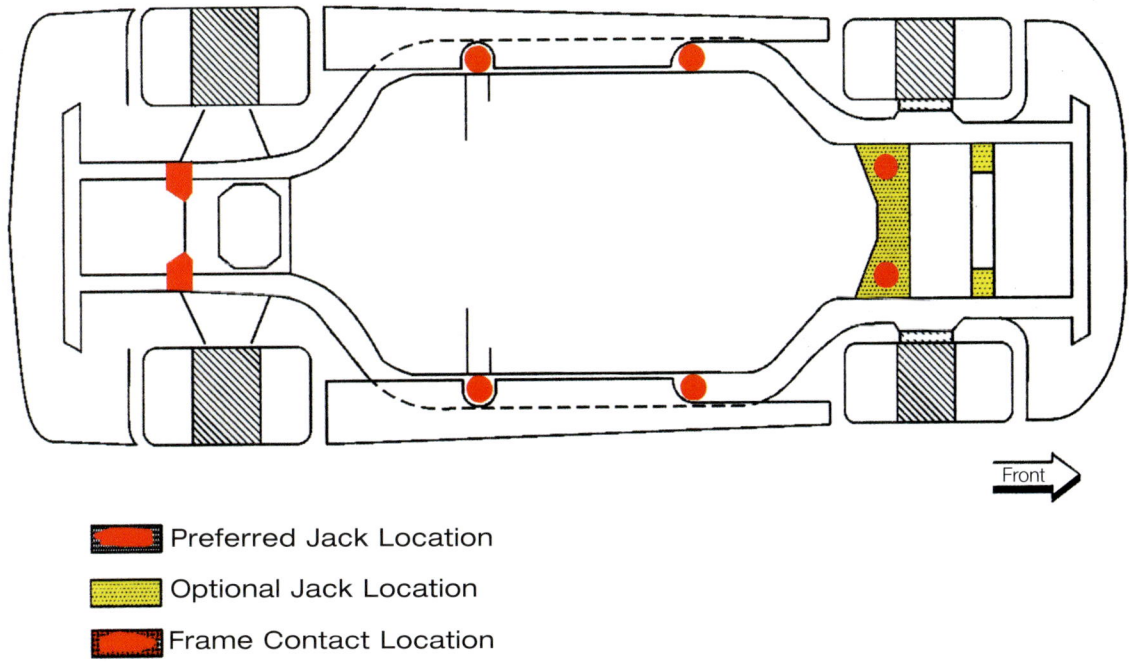

- ■ Preferred Jack Location
- ■ Optional Jack Location
- ■ Frame Contact Location

These are the official jacking points. *Courtesy of General Motors*

in front of each tire. Make sure that these pieces of wood won't slide when you drive the car up on them. You might try gluing sandpaper to the bottom of the boards to create friction between the board and garage floor, thus holding the board in place. You may, however, end up with sandpaper marks on your garage floor. Be careful.

Slowly drive the car onto the wooden boards. You might need to give your car a small "running start" to get on the blocks. If you hit them and the car doesn't go up on them, back up and try it again. If you force the issue, you could flip the wood planks up with enough energy to hit and possibly crack the fiberglass underneath.

Once the car is on the blocks, make sure that you've engaged the parking brake and that the transmission is in park. Also, if you're going to be removing your wheels, make sure you loosen the wheel nuts before lifting the car.

Once the car is safely on the wooden blocks, take a floor jack to the front of the car and center a piece of 2x6x25-inch wood on the jack. Move the jack and the 2x6 under the front end. Pivot the lower air dam out of the way to move the floor jack and wood under the air dam.

Now, position the jack and the wood under the front aluminum crossmember, being careful to not let anything touch the fiberglass leaf spring. Raise the floor jack until the wood makes contact with the crossmember, making sure that both the wood and the floor jack pad are centered under the crossmember.

After you've lifted the car to the required height, move the two jack stands under each end of the wood.

The fiberglass panels wrap right around the frame rail, which is wonderful until you crunch your jack into the fiberglass.

Leave enough space in the middle to remove the jack from under the car.

Now move your jack to the rear of the car and repeat the same procedure. Raise the floor jack with the other 2x6x25-inch piece of wood under the rear crossmember, again being careful to not let anything touch the rear fiberglass leaf spring. Place the two jack stands under the car the same way you did on the front and remove the jack.

The rocker panel rolls under the car. You can also see the rather minuscule cutout that was designed for lifting your Corvette. It's too easy to destroy several hundred dollars worth of bodywork on your Corvette when lifting the car from the side. I prefer using a jack on the front or rear of the chassis.

With the four jack stands in place, the car should be completely off of the ground. There will be no metal touching the aluminum crossmembers and the car can be left like this for days if needed.

If you're going to lift the rear only, you still need to drive the car onto all four wooden planks. If you don't do this, you can accidentally lift the rear high enough to cause the front fascia to touch the ground. Be careful.

These plastic plugs snap into the frame rails and provide nice jacking points. You can either leave them in place all the time, or install them before you have your Corvette serviced. I would also leave a note in the car that explains that the service technicians should use these pads as jacking points. Then before you leave the service facility, go around your car and make sure that none of the lower panels are cracked.

Jacking Pucks

If you look at the bottom of each frame rail, just where it turns inboard to clear the wheel, you'll see a welded-in (1997 and early 1998) or riveted-on (later 1998 and on) circle of steel with a slot in it. These shipping slot reinforcements are used to hook the car onto the truck that transports it to your local dealer.

These reinforcements are a great place for the jack stands' support pads. The problem is determining how to get the car high enough to place the jack stands on these points.

There are small plastic jacking pucks that can be placed into these slots so that they extend down below the body panels. The jacking pucks are designed to fit into the transportation slots and, with a simple twist, lock into place. Most Corvette catalogs list these pucks.

Using a Lift

Most shops today use frame-contact hoists with arms that come in from the side. Because the rocker panel cutouts aren't large enough, lifting the C5 Corvette with the pads found on most hoists is impossible.

You will need a pair of wooden blocks small enough to fit inside the cutouts. It's easiest to simply cut them out of 1-inch-thick blocks of hardwood, 2 to 2 ½ inches on a side. Position these pads in such a way that the blocks contact the shipping reinforcements but not the rocker panel material. Then, raise the car—carefully.

If you have a set of plastic jacking pucks, you might consider simply inserting them into the frame rails before you take your car into the shop to prevent any possible problems.

PROJECT 2 | Changing the Engine Oil

Time: 30 minutes

Tools: Metric sockets (⅜-inch drive) and metric combination wrenches

Torque: Oil drain plug: 18 ft-lb

Talent: ★★   **Applicable Years:** 1997–2004   **Tab:** $30

Tip: Place newspapers under the drain plug and oil filter, since some oil will run down onto your garage floor.

Performance Gain: If you decide to use 20W-50 oil, keep in mind that you'll see an increase in your oil temperature. It's best to stick to the recommended 5W-30 weight oil.

Complementary Work: Lube the chassis every time you change the oil. Also, take a few extra minutes to oil the door hinges and hood latches.

Changing the oil is very easy with the 1997–2004 Corvette. Your local discount-parts house has everything you need. I use the ACDelco oil filter and Mobil 1 oil, but the brand of oil and oil filter are subject to personal preference.

I remember an excellent oil filter study done over a decade ago under highly controlled conditions. Interestingly, some of the filters were found to be filtering too well, catching every little microscopic element. This meant that within 1,000 miles the filter material was totally plugged. Beyond 1,000 miles, the oil was being sent through the bypass circuit and not getting *any* filtration. Today, the average filter is designed to catch particles in the range of 25 to 30 microns. The average human hair is usually 67 microns in diameter.

The single most important part of an oil filter is the gasket. If the gasket fails, you'll lose the oil and that will be the end of one very nice Corvette engine. The cheap oil filter manufacturers use a lower-quality material in the construction of this base gasket—one reason to stick to a quality filter brand.

General Motors recommends Mobil 1 synthetic oil in its Corvette engines, and that's good enough for me. Everyone seems to have an opinion on which oil is best, but it's generally an uninformed opinion. While there are likely differences between engine oil brands, a friend of mine once observed: "I've never seen an engine explode because of the brand of oil. I've seen a lot of them, though, that had serious problems because the oil was low or never changed."

You should be aware, however, that in 2004 GM began using oil with low phosphorus content. Beginning in 2004, all catalytic converters are under warranty for 120,000 miles. The active ingredients in a catalytic converter are platinum, rhodium, and palladium—all very expensive metals.

Poisons in the exhaust gases cause most catalytic converter failures. Traditional motor oils contain both phosphorous

The oil filler cap, which no one ever reads, has good information about the type of oil you should put in your Corvette.

and sulfur. Rather than increasing the amount of expensive metal used in the converters, the manufacturers decided to have the oil companies remove as much sulfur and phosphorous from the oil as possible. That way, they could get the oil companies to absorb the increased cost for the tighter emissions standards and the automobile companies could hold down the manufacturing cost of the catalytic converters. In order to remove the sulfur, the oil companies had to develop a new refining process.

There's a misconception that you need to use heavy oil in an older or high-mileage car. Older cars really don't need heavier oil, and there is no benefit to running it in your Corvette. The 20W-50 oil simply gets hotter than thinner oil. If your Corvette is running over 200 degrees all the time, you should use 5W-30 or 10W-30.

BASICS

Oil filter gaskets are critical. The better oil filters use a superior gasket. When you remove the old oil filter make certain that the old gasket comes off with the filter rather than sticking to the block surface. Also, make sure you lube the new gasket before you put the filter in place. I generally fill the filter halfway with new oil before screwing it into place.

This is the original C5 oil pan. The huge bat wings solve an oiling problem that the LS1 engine has in sustained cornering, but it takes forever for the oil to drain out of these side wings. Raise the rear of the car slightly higher than the front and then remove the drain plug in the front of the pan. When you come back several hours later most of the oil will have drained out.

GM seems to have finally solved the oiling problems with the C6 Corvette. The C6 returned to a rather conventional cast-aluminum oil pan, making it a lot quicker to change the oil.

In the case of the C5 Corvette, the factory recommends changing the oil every 7,500 miles, or when the change-oil light goes on. The 3,000-mile oil change is an idea that has been promulgated by the oil companies. If they can convince you to change your oil twice as often as necessary, they get to sell twice as much oil.

Since 1990, Corvettes have been equipped with an engine oil life monitor that records the number of engine starts, number of engine revolutions, coolant and oil temperatures, and other data. It displays the oil life remaining and will turn on the change-oil light when the programming determines that the life of the oil has been reached and the additive package depleted.

The C5's oil life monitor does not analyze the oil. It makes an educated guess at its condition by measuring trends in engine operation. General Motors recommends that you change the oil and filter between 7,500 to 10,000 miles of driving. If you don't drive your Corvette that much, simply change the oil once a year.

The Oil Change Light

The C5 Corvettes are equipped with an engine oil life monitor that uses the central control module (CCM) to calculate the lifespan of the engine oil by measuring engine rpm and oil temperatures over the time you drive the car. The lifespan is calculated based on certain things that happen to the car, rather than monitoring the actual condition of the oil like some European systems. Reset the engine oil life monitor after every oil change.

Resetting the Oil Life Monitor

(1) Turn the ignition on, but do not start the engine.
(2) Press the "reset" button to clear any messages showing on the DIC.
(3) Press the "trip" button until the screen reads "oil life remaining XX%."
(4) Press and hold the "reset" button until it resets.
(5) Turn the ignition off.

The next time you turn the car on, the oil life monitor will be at 100 percent and will shortly change to 99 percent. After you do this, you can also reset the "Trip B" odometer so that you'll know how many miles are on that oil at any given point.

The C4 engine had some oiling issues under hard and prolonged cornering. GM decided to counter the problem on the C5 with a "batwing" oil pan design. Because it takes a while for the oil to drain from the wings, you shouldn't take your C5 Corvette to a speedy oil change shop that might try to rush the draining process.

The C5's drain plug is located at the front of the oil pan. The Corvette engine holds 6 quarts; make sure your drain pan will hold 6 quarts of oil as well. Remove the plug and allow the oil to drain for 15 to 30 minutes. Once the oil is drained, wipe the drain plug and its sealing surface on the pan with a clean rag, replace the drain plug, and tighten it to 18 ft-lb.

When you change the oil in the C5, the rear end should be just a bit higher than the front. You can do this by setting the rear jack stands one hole or notch higher than the front, or just raise the rear slightly with your floor jack while allowing the oil to drain out. This is especially necessary on early C5s, which have large cast-aluminum oil pans. This oil pan style uses the so-called "bat wing" extensions. These oil pans take forever to drain. The best way is to raise the rear of the car, remove the drain plug, and go to lunch. By the time you return, the oil pan might be fully drained.

The oil filter is partially recessed into the oil pan casting. Remove the oil filter with a cap-type oil filter wrench or a strap-type wrench with a swivel handle.

Make sure you wipe the gasket surface on the engine clean. Then fill the new filter about halfway with engine oil, put some oil on the gasket, and screw the filter onto the block.

After the gasket contacts the filter mount, tighten the filter ¾ to 1 turn farther. Finally, add 6 quarts of new oil and replace the fill cap. Set the parking brake, start the engine, run it for about three minutes, and then shut it off.

Crawl back underneath your Corvette and inspect the filter and drain plug area for leaks. Wait a few minutes, and then check the oil level. If necessary, add oil to bring the level in the crankcase up to the "full" mark on the dip stick.

Heat causes the oil to break down. Keep an eye on your Corvette's oil temperature gauge. If for any reason your oil gets into an abnormally high range, you should change it at the earliest possible opportunity.

More commonly, your engine may overheat from a burst hose or low coolant levels. If you've overheated from these problems, change your oil at the same time you make the other repairs. Don't take a chance on using cooked oil in your Corvette.

Dispose of your used oil and filter in an appropriate fashion. The Environmental Protection Agency (EPA) calculates that about 75 percent of oil filters, a hidden source of both used oil and recyclable steel, end up in landfills. Most major parts stores have programs in place to help you dispose of used oil and oil filters.

PROJECT 3 | Replacing the Serpentine Belts

Time: 30 to 60 minutes

Tools: ½-inch breaker bar and metric sockets

Torque: Belt tensioner to engine block: 30 ft-lb

Talent: ★★   **Applicable Years:** 1997–2004   **Tab:** Under $50

Tip: Check both of these belts on a regular basis for signs of wear.

Performance Gain: There would be a performance gain only if the pulleys were changed, and that's usually not a good idea.

Complementary Work: Replace the serpentine belt tensioners if you've been hearing a noise coming from the belt area.

BASICS

C5 Corvettes use the customary serpentine belt system. There's one belt for the air conditioning compressor and another for the alternator and water pump, power steering, and other accessories.

Plan to replace both of these belts around 30,000 miles. The main causes of wear on the belts are heat and mileage. Every time the belts pass around a pulley, they bend and flex, producing heat that hardens the rubber over time. After millions of journeys around the pulleys, even the best drive belt begins to suffer the effects of age. The rubber begins to crack and fray, and the internal cords become weak and brittle.

Examine the belts for cracks and frayed edges. Some small cracking is normal. If you find there are more than 8 or 10 cracks per inch, it's time to replace the belt. If there are chunks of rubber missing from your belt, you've been driving on borrowed time. Replace the belts as soon as possible.

Frayed edges are another sign that you should think about replacing the belt. Frayed edges are not a normal wear pattern and may indicate another problem. Check to see if all the pulleys are in alignment.

I generally avoid parts catalogs and take the old belt to the parts store with me when I purchase a new belt. This way I know I'm getting the correct length the first time. If you think you may have trouble remembering which way the belt goes around the pulleys, take a digital picture or make a drawing. If you rely on your memory alone, it's guaranteed that a

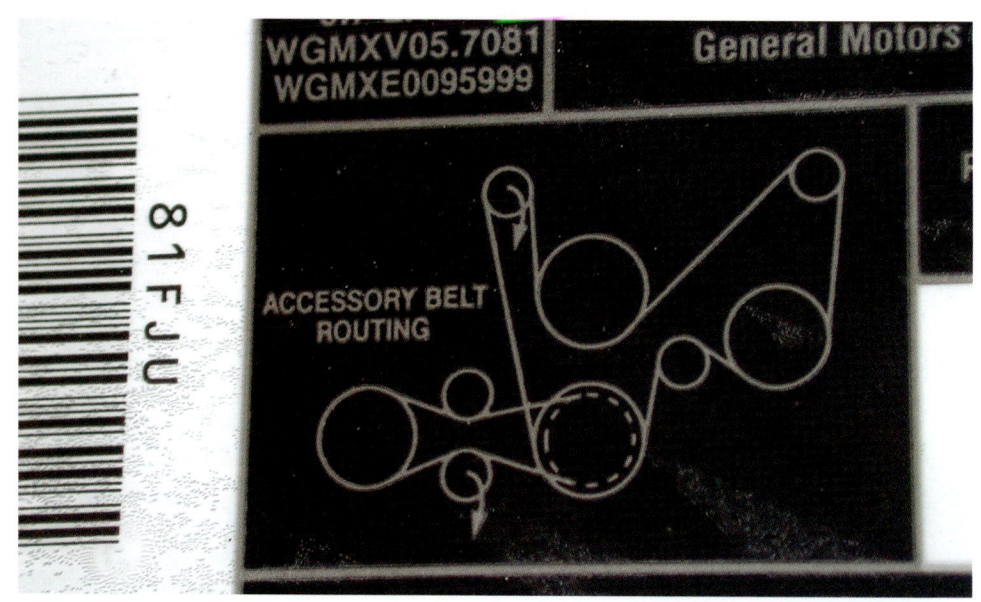

The C5 uses two serpentine belts. They're extremely easy to change and not very expensive. You should be able to find a label that shows the routing of the belts, but I've noticed that a lot of C5s don't have this label. In that case, make a drawing or take a picture before you remove the belt.

24

The idler pulleys seldom go bad. When there is a problem, though, they make a chirping sound that is very similar to the noise the old V-belts used to make.

family crisis will erupt in the middle of this job. Write the belt path down on paper.

Noise Problems

Sometimes you'll hear a chirping sound from the area of the belt. This could be a problem with the alignment of the pulleys. Using a spray bottle filled with water, lightly mist the belt. If the noise level recedes for several seconds but then returns, you have a misalignment problem. If the noise immediately increases after the belt is sprayed, slipping is likely the problem. If the water-spray test is not conclusive in diagnosing the problem, remove the belt and reinstall it so that the belt runs in the opposite direction. Because misalignment noise is influenced by the direction of misalignment in the drive, flipping the belt around in this manner will eliminate or significantly diminish (temporarily) any noise caused

The belt system on the C5 is wonderful, and you can gain a little performance by swapping over to a set of aftermarket pulleys. *Courtesy of General Motors*

25

You don't have to have a special tool to move the belt tensioner. In fact, a 15-millimeter box-end wrench works nicely.

BASICS

by drive misalignment. If the noise remains unchanged, the problem is not likely related to drive alignment.

Serpentine Belt Tensioners
This chirping sound usually isn't coming from the belt, but rather from one of the tensioners. These tensioners are designed to keep constant tension on the serpentine belts.

With the engine turned off, look at the tensioners. A drive belt tensioner is a spring-loaded device that sets and maintains the drive belt tension. The drive belt should not require tension adjustment over the life of the drive belt. Automatic drive belt tensioners have drive belt wear indicator marks. If the indicator mark is not between the MIN and MAX marks, the drive belt is worn and both the tensioner and belt should be replaced.

Also note that when the engine is running, the belt tensioner arm will move. Don't replace the belt tensioner simply because of movement in the belt tensioner arm—it's supposed to do that.

If you still think you have a tensioner problem, remove the drive belt. Then use your 15-millimeter box-end wrench on the belt tensioner pulley bolt and move the belt tensioner through its full travel. The movement should feel smooth, and there should be no binding. Also, the belt tensioner should return freely. If you have any binding as it returns to the relaxed position, replace the drive belt tensioner. Remember, though, that if you allow the drive belt tensioner to snap back into the free position, you run the risk of damaging the tensioner.

The next trick is to remove the belts from the engine and start the car. Then listen for the noise that's been making you crazy. If the noise isn't present after removing the belts, then the problem is either in the belts or the belt tensioners. You can even do this trick with one belt at a time to further isolate the noise. If you decide to use this trick, do it quickly, since the water pump won't be circulating coolant with the belt off. If you take too long, you can overheat the coolant. You'll also probably set off a trouble code by running the engine with no belts, but that can be easily erased when you're all done.

Changing Pulley Size
Several companies sell different-size pulleys that are supposed to give you more horsepower. What they usually give you are more problems. All of the drive pulleys on the C5 Corvette are carefully calibrated. The water pump and the alternator are designed to run at specific speeds. GM engineering looked at how most people drive their cars and then calculated how fast these items should revolve. If you start changing the drive pulleys, be prepared to have coolant temperature problems and a dead battery every now and then. True, you'll get a couple more horsepower at 5,000 rpm, but you'll pay a price for this. Changing pulley size is like messing with nature—it can be done, but the results aren't always pleasant. And at the end of the day, any performance gain is truly minimal.

PROJECT 4

Replacing the Air Filter

Time: 15 minutes

Tools: A screwdriver to pop the clamps **Torque:** N/A

Talent: ★   **Applicable Years:** 1997–2004   **Tab:** $15 to $145

Tip: Make sure that the air filter lid is square on the housing before you tighten everything down.

Performance Gain: There's very little performance gain in the various air filter brands.

Complementary Work: Clean the throttle body intake passage with air intake cleaner.

The first aftermarket air intake systems appeared within 30 days of the C5 Corvette's introduction. Before you run out and purchase one, keep in mind that GM designed the air intake system with several things in mind.

The goal of an air intake system is to get as much air as necessary into the intake runners. An engine requires only so much air; it isn't necessary to jam massive amounts of air into an engine if the exhaust system can't get rid of it quickly. The opening and closing of the valves, along with the size of the intake runners, determines how much air the engine needs to digest.

For that reason, you won't get much horsepower from simply adding a bigger air filter. Unless you've ported the cylinder heads, changed the camshaft, and replaced the exhaust system, you don't need an aftermarket air filter system. The engine is already getting all the air it needs.

General Motors had trouble with water ingested into the C4 Corvette's intake system. GM engineers were diligent to avoid the same problem on the C5.

Dave Hill, the Corvette's chief engineer, has always had a thing about noise, and he was adamant that intake noise in the C5 should be minimal. This is where the aftermarket sys-

This air induction system is designed for a supercharged car. The intake pipes run down and pick up air from the two front brake ducts.

BASICS

All of the aftermarket induction systems pick up air in the same place—in front of the radiator. This means that they're all picking up cool air. The differences between the various brands are marginal at best. Most of the aftermarket brands have a much higher noise level than the stock version. This means that when you accelerate rapidly, you'll think you're going a lot faster than you are. It does sound nice, though.

BASICS

There are three different air filter lids for the C5 Corvette. The best one is the later Z06 lid. This is perhaps one of the most cost-effective induction changes you can make.

tems went in the other direction. Air intake noise can give you a real feeling of power. You may not even notice where the noise is coming from, but suddenly the car feels more powerful. The aftermarket people know that you'll probably never use a dynamometer to measure air filter effectiveness but that you do want to feel like you're going fast. More intake noise will give you that feeling.

Corvette C5 air filters are very easy to change. Simply loosen the clamps and remove the lid on the intake housing.

You don't even need tools to do this, although a long-bladed screwdriver or pliers will sometimes help.

Be careful, however, with the soft urethane sealing gasket molded in place on the filter element. Make sure that the gasket is seated carefully in the housing before you tighten everything down. If the filter isn't properly seated, it'll allow unfiltered air into the engine.

The C5 Corvettes all use what's called a linear-flow air filter. This means the air passes directly through the air filter and doesn't have to turn any corners. This air path produces minimum airflow restriction.

The stock ACDelco air filter is already a good air filter, and a change to a different brand will result in only minimal, if any, power gains. There's little to be gained from a switch to a K&N or Accel air filter. They'll last as long as you own the car, but they cost more than twice as much as the stock filter.

If you're running your Corvette at the drag strip or a road racetrack, you'll pick up a few horsepower (maybe two or three) with a modified air filter lid and a K&N air filter, but only at the highest rpm. You'll lose horsepower with a dirty or nonexistent air filter.

Some aftermarket companies take two filters that meet in a "Y" before entering the intake plenum. This seemed like a great idea until people found that the intersection of the two air paths created tremendous turbulence right at the juncture. The result was that less air was entering the plenum than with a single filter.

PROJECT 5 | Replacing the Fuel Filter

Time: 30 minutes

Tools: 5/16-inch fuel line disconnect tool   **Torque:** N/A

Talent: ★★   **Applicable Years:** 1997–2003   **Tab:** $25

Tip: The fuel line disconnect tool makes things a littler easier, even though most professionals don't use one.

Performance Gain: N/A

Complementary Work: While this can easily be a stand-alone job, it's often combined with an air filter and oil change.

Corvette fuel filters are extremely easy to change. If you haven't done it before, the only problem may be actually finding the filter. Be sure to change the filter in a well-ventilated area, since you'll likely spill gas on the floor.

Fuel filters became critical once the change to fuel injection was complete. The fuel passages in an injector make a carburetor look like an expressway. Given the inexpensive price of a Corvette fuel filter, you should change it every year, maybe more often if the spirit moves you. Although, in the middle of the 2003 production run, General Motors declared the fuel filter to be a lifetime component.

Your Corvette has one of three different fuel filters, depending on what type of fuel system you have. If you have a single stainless-steel braided fuel line going to the engine, then you have the GM 10299146 fuel filter, which has a fuel pressure regulator built inside of it. If you have two stainless-steel braided fuel lines, then GM 10287788 is your filter. This filter doesn't have a pressure regulator built in.

The third type—installed beginning in the middle of the production run for 2003—is called an FFS fuel system and has the fuel filter in the tank. It's not necessary to change this filter. On the early Z06 Corvette C5s, the fuel filter was

There is a tool designed to remove the clips that hold the fuel filter in place, although most people simply use their fingers as a tool. You can get the special filter tool at most NAPA stores.

BASICS

29

To remove the filter, you have to first remove the bracket and then compress the little plastic tabs that hold the filter to the fuel lines.

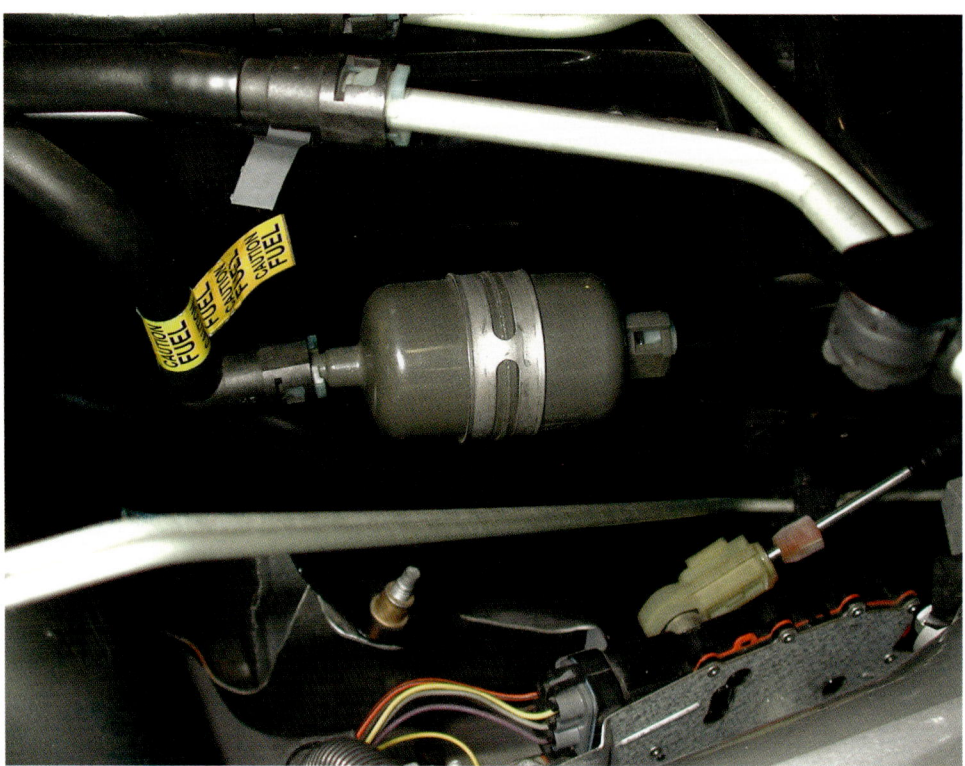

located on the inside of the driver's side frame rail, just in front of the rear wheel. Later Z06s have the FFS fuel system and filter.

There are a lot of differences between the FFS system cars and non-FFS system cars. The tanks are connected in a different way, and it's not as easy to drop the tank as it was with the earlier cars. But according to GM, you will not need to replace this fuel filter as long as you own the car.

Always release the pressure in the fuel system before you exchange the old filter with a new one. You'll find a Schrader valve located on the fuel rail in the engine compartment. This is the valve that looks just like a tire valve. If you push down on the valve, the same way you would let air out of a tire, you can release the pressure in the fuel system. But don't do this on a hot engine, since you could end up burning down the car right in your driveway—just try explaining that one to your spouse.

With the pressure in the fuel system released, jack the car up on the left side and place two jack stands under the frame. The rear jack stand position may require some thought since you don't want it to interfere with your effort to remove the fuel filter. Another way is to simply raise your Corvette by the rear crossmember and place the jack stands on the rear crossmember. You have choices here.

If you haven't released the fuel pressure in the system, be prepared for a shot of gasoline when you remove the line from the filter. Sometimes I cover the whole thing with a shop rag so I won't get gasoline in my face. Wear safety glasses, just in case.

The first trick is simply to locate the fuel filter. The filter is under the frame in the rear of the driver's side rear wheel. First, remove the bolt holding the filter assembly to the frame. Then comes the messy part—to remove the two quick-connect fittings toward the rear of the filter, you should be able to just squeeze the blue tabs and pull the two lines from the rear side of the filter. If you're using a fuel line disconnect tool, remove the line on the front side of the filter first. Again, some gas will pour out, as the filter is full of gasoline. The $5/16$-inch fuel line disconnect tool is made of metal or plastic and looks like a ring with a cutout and a tail that slides between the connector on the line and the hard tab on the filter. Don't be right under this fuel filter, since a couple of cups of gas will come pouring out.

Now clean up all of the fuel on your garage floor or driveway. Put the new filter in and bolt it back in place. Reconnect the lines. Turn the key in the ignition to pressurize the fuel system and fuel lines, but don't start the engine. Turn the key two or three times, and then crawl back under the car to check for leaks.

Once the new filter is in place (and not leaking), you can remove the jack stands and start the engine. Let it run for a couple of minutes, and then turn it off. Now poke your head under the car and check once more for fuel leaks. Anytime you replace a fuel filter, you need to make absolutely certain that all the connections are tight. A leaking fuel filter is a major cause of car fires.

PROJECT 6 — Detailing Your Corvette

Time: Several hours

Tools: N/A   **Torque:** N/A

Talent: ★★   **Applicable Years:** 1997–2004   **Tab:** Under $50

Tip: Always use the least-aggressive product that will restore the finish. Never start with the industrial-strength product. Regular maintenance will ensure that these milder products get the job done.

Performance Gain: Your Corvette looks better, which means it feels faster.

Complementary Work: This is a good time to check your oil level and tire pressures.

More time is spent detailing the Corvette than all the other things we might consider. We can talk about new computer chips and exhaust systems at great length, but wax is the one subject that is near and dear to every Corvette owner.

When you look at your C5 Corvette from certain angles, the surface resembles the skin of an orange. GM isn't the only company to have problems with "orange peel" in the paint, but that doesn't make you feel any better. The only real cure is to have the car repainted. But a good detailing with quality wax will minimize the orange-peel effects. When you wash your car, make sure you don't use dish soap, which will remove the wax from previous efforts.

Detailing Clay

Once you've gotten all the dirt off the car, you can use detailing clay, but do it inside the garage to keep dust particles from floating around. Detailing clay is a great invention, originally designed as a way for body shops to remove overspray from repaired cars. It acts as a very-fine abrasive that removes any surface blemishes from the paint. It doesn't actually remove any paint, just debris.

Use a lot of lubricant, such as Meguiar's M-3416 Final Inspection, and then hydroplane the clay across the surface to abrade any little bumps from the existing paint. You can float this clay back and forth across the paint until the resistance ceases, indicating that the unwanted droplets—often tree sap and insect remnants rather than overspray—have been removed from the surface. Clay smoothes the surface of the paint by removing the microscopic blemishes much like sandpaper smoothes a piece of wood. These detailing clays are the best way to remove surface contaminants.

If you're not careful, though, detailing clay could be a scratch waiting to happen. Use detailing clay only on well-lubricated areas, and constantly check your clay for contaminants. If you rub the clay on areas of the paint that haven't been well lubricated or a piece of grit lodges in the clay, you've just created sandpaper. You should constantly refold the clay to expose a fresh, clean surface. If contaminants lodge in the clay, simply tear off the section and throw it out. As soon as you're done using the clay, place it in a sealed plastic bag. Experiment with different brands of detailing clay. I've found some to be very effective, while others are a waste of effort.

Wax

Carnauba wax is very popular but isn't as effective as synthetics. The paint on your Corvette is a miracle of chemistry. And, from a chemist's point of view, synthetic waxes are closer in substance to the Corvette's paint than a wax produced by nature.

Zaino Show Car Polish and Liqui-Tech's Finish First, both synthetic products, will hold up better than a carnauba-based

I generally apply detailing spray to the whole car every time I wash it. I use it on all the window glass to remove the water stains that remain after drying.

The best wheel cleaner you can use is car-wash soap. Use the least aggressive product that is still effective. Plain car wash will cause few problems and will generally remove all the brake dust from your wheels. If you let them collect too much dirt, you may have to go with a more aggressive wheel cleaning product.

wax. The range of durability depends on a lot of factors, but the broad statement is still true. Your geographic location will have a greater effect on the staying power of your wax than the specific product. Most of the chemists in the detailing industry give the edge to a polymer wax over a carnauba wax every time.

The Interior
Even though the C5 Corvette has leather seats, there's a tremendous amount of vinyl and plastic in the car. Vinyl is a raw semi-liquid that is held in place by a solid skin. Ultraviolet (UV) rays constantly bombard the dash and other vinyl parts. These UV rays break down the molecules in the skin, allowing the raw vinyl to escape. This is referred to as "off-gassing." This vinyl deposits itself on the glass, forming a haze.

Many vinyl cleaner/dressing products don't contain UV protectants, and the silicone oils in some products may act as a magnifying glass, actually intensifying the UV degradation. Silicone oil may also dissolve the oils in the vinyl skin, hastening the premature cracking.

A quality vinyl cleaner/dressing product will contain a UV protectant. Allow it to work into the surface for a few moments and then buff off the excess. Treat the top of the dash more often than any other area, since it's subject to the most severe attacks by UV rays and heat.

My favorite vinyl treatment is from Just Dashes (www.justdashes.com). Their main business is the reproduction of dash panels. Since they spend so much time making some of the best dashes in the hobby, it's only natural that they would help owners keep the dash looking good.

Leather
I have reservations about most of the products on the market to treat leather seats. Lexol is the number one product on the market, followed closely by Duragloss LC (leather conditioner). Lexol and Duragloss are the better choices from a bunch of bad choices available.

Whatever product you decide to use, try it on a hidden part of your leather seat. Make sure that you like what happens before you treat both seats in your Corvette.

If you're not sure how often to treat the seats and the steering wheel in your C5 Corvette, pay attention to the product—it will let you know. The goal in treating leather surfaces is to replace oils. You'll notice how quickly the Lexol soaks into the leather. If you treat too often, the Lexol will simply build up on the seat. If you wait too long, the Lexol will soak quickly into the leather.

The leather seats in the 1997–2004 Corvettes are very durable. The stitching usually gives out before the leather goes bad. I've had my seats restitched through the original holes several times now. It usually runs about $25 a seat if I take the seats off the frames first. I suspect a lot of the leather conditioners actually weaken the stitching, so don't get too carried away with treatments.

Wheels
The biggest problem with wheels is brake dust. The metallic dust, given off during braking, is red-hot and will actually burn tiny holes in the finish of your wheels. If you keep up with removing brake dust accumulation, a simple car wash solution is enough to clean your wheels.

If you have small droplets that look like road tar on your wheels, they may in fact be repolymerized brake pad adhesive. Brake pad adhesive is the root of most of our wheel problems.

Don't let your leather seats dry out. GM (and Lear, who manufactured the seats) used really cheap leather in the C5. Regular treatment with a conditioner is essential to preserve the seats. Remember to clean all of your leather before you apply the conditioner.

This is still my favorite vinyl product. Just Dashes (www.justdashes.com) is the leading expert in dash panel restoration. This product was designed to make their restorations look great after installation. It works great for your C5 interior and avoids that glossy look you get with most vinyl products. It also holds up nicely, which means you don't have to make detailing the vinyl a weekly chore.

The polymer adhesives form droplets that wind up on the wheels, where they adhere with a vengeance. When this adhesive residue becomes wet, it turns acidic and will actually etch your wheels. The only solution is to clean your wheels often.

All of the stock wheels for the 1997–2004 Corvettes have either a painted surface or a clear coat. You can treat these wheels the same way you handle the painted body. If you clean your car often enough, just use normal soap and clean the wheels with a separate rag. It's best to clean them when they're cool.

P21S (www.p21s.com) is the only wheel cleaner I use on my wheels. Many of the popular brands are highly acidic and can do more harm than good. The active ingredient in a lot of wheel cleaners is hydrofluoric acid—the same substance used to etch glass. P21S may not be as aggressive as other brands, but it will not strip the finish off your wheels.

Most wheel cleaners work best on a dry wheel. Spray the cleaner on the wheel and work evenly into all areas of the wheel with a soft cloth or a small sponge. Allow the wheel cleaner some time to work (3–5 minutes) and then gently scrub the wheel with a wet cloth or sponge.

If you have the stock Corvette wheels, consider waxing them with a synthetic wax. The damaging effects of red-hot brake dust, brake dust acids, pollution, and ozone will be unleashed on the wax and not your wheel.

If the wheels are slightly faded or dull-looking, 3M Imperial Hand Glaze may help. Apply the glaze to a soft cloth, gently rub out the clouding, and buff out. If this doesn't do the trick, put a generous amount of 3M on your cloth and add a dime-size amount of P21S Metal Finish Restorer Metal Polish. Polish out the clouding with this combination and follow up with a coat of wax. Treat your wheels like you would any other painted surface on your Corvette.

If you have polished aluminum wheels, they will take longer to detail than the rest of the car. Stick with Simichrome polish for aluminum wheels.

Tires

Once again, I'm back to an old product after messing around with several other products on the market. I gave up on One Touch tire dressing years ago because the look was too glossy and the results always looked streaked.

Then I decided to try One Touch with one of the new sponge applicators that can be found at your local parts discounter. It worked so well that I use One Touch all the time now.

I also recommend tire-care products from Stoner. This company got started making release agents for the tire molds at Goodyear. Stoner products are easy to apply, since they come in aerosol cans and all you have to do is spray. One Touch holds up a little better, but Stoner products give you a little less gloss.

Some tire-care products use raw silicone, which may actually lead to sidewall cracking over time. Some tire treatments also contain formaldehyde, but the manufacturers are not required to include that information on the product label.

Carpet

Carpet is carpet. Anything you use in your home is good for your Corvette. If you vacuum the inside of your Corvette on a regular basis, you can probably avoid the use of any carpet cleaners. When it comes to carpet cleaners, the least-aggressive product is the best.

When it comes to tire gloss, I suggest that you visit the local discount parts house and buy four different brands. Use one brand on each tire and then decide which you like best. Don't forget to see if the product rubs off onto your slacks the next day. Some tire treatment products can leave nasty stains several days after application.

SECTION TWO
ENGINE

Projects 7–17

PROJECT 7	Cleaning the Throttle Body

Time: Less than two hours

Tools: Throttle body; flat-blade screwdriver; ⅜-inch drive ratchet; 3 ⅜-inch drive extension; 10-millimeter, ⅜-inch drive socket; needle-nose pliers; torque wrench capable of measuring 9 ft-lb

Torque: Throttle body bolts: 9 ft-lb

Talent: ★★   **Applicable Years:** 1997–2004   **Tab:** Less than $50

Tip: Always use air intake cleaner, not carburetor cleaner.

Performance Gain: A smoother idle and improved low-speed performance.

Complementary Work: Always use a new air filter when you are done cleaning the throttle body.

The throttle body is an often-ignored engine component. We just assume all the air gets to the combustion chamber. We know it has to get through the air filter, but we seldom think about the throttle body. The throttle body on your Corvette gets just as dirty as an old carburetor and can really benefit from a good annual cleaning.

Remember that the throttle body is a system designed to control dry air. Until the air reaches the fuel injectors, there is simply a dry air passage. (Those of us raised on carburetors have a tendency to forget that simple fact.)

Some of the C5 throttle bodies are simply worn out, and a lot of them are incredibly dirty. This filth can cause a variety

The inside of the throttle body gets incredibly dirty after a few years. You should clean it on an annual basis. If you're having an idle problem, this is the first place to look for a solution.

34

Air Intake Cleaner is far superior to any of the old carburetor cleaners we used to use. It's also becoming easier to find at auto parts stores.

of problems, the most common being a fairly high or erratic idle speed, caused when the engine vacuum can't completely close the throttle plates.

There are several approaches to cleaning the throttle body, but the best is to keep everything clean on a regular basis. You could probably do it on a bimonthly basis. Just remove the air intake assembly and spray some air intake cleaner on the throttle plates.

If your Corvette is a little older, chances are that no one has ever cleaned the throttle plates, so you'll have to remove the throttle body from the car. Fortunately, it's not that hard, and once you thoroughly clean it, you can maintain it by cleaning it annually.

Procedure:
Disconnect the red positive terminal on the battery before you do anything else.

Loosen the screw clamp that holds the air bridge to the throttle body. It's easiest to remove the entire intake duct from the car. This means loosening the clamp from the other end as well.

Unplug the wire that connects to the throttle position sensor (TPS), located on the passenger side of the throttle body. Be careful when you disconnect the electrical connectors that you don't break the locking tabs. Since we're also going to disconnect a coolant hose, make sure this plug is positioned so it won't get wet from the coolant.

Now unplug the wiring connector that goes to the electric motor on the driver's side of the throttle body. Use a small screwdriver to depress the connector tab, and gently wiggle the connector loose.

Using needle-nose pliers, compress the small gray spring clamp that holds the coolant return line in place. Move the clamp approximately 2 inches down the line, but don't remove the hose yet.

Remove the coolant overflow tank cap, located on the passenger-side fender well with a yellow sticker on the cap. Remove the cap very slowly, allowing the pressure to escape the tank. Leave the cap off for now.

Find a ¼-inch rubber cap (vacuum line caps are good for this), and remove the rubber hose from the end of the hard line. Place the rubber cap over the end of the line to prevent spillage. Keep the rubber hose in the upright position and place the hose against the air conditioning lines on the passenger-side fender well. Keep it pointed up to prevent spillage.

Now remove the PCV return line from the throttle body. Place it aside and out of the way.

Use a 10-millimeter socket to remove the three bolts that hold the throttle body to the intake manifold. Gently move

If you begin to modify your engine for improved performance, you'll need a TPIS (www.tpis.com) throttle body. By itself it won't do a great deal for you, since it should always be part of a total performance program.

The TPIS throttle body is truly a bolt-on addition. Remove the old throttle body and simply bolt the new TPIS unit into place. A little time on the dyno will help maximize performance and drivability.

This bypass tube is used to eliminate the hot coolant passages that run through the throttle body. If you live in a warm climate, this is a modification to consider. Make sure that the one you purchase has a barbed end and isn't just a piece of plain tubing.

Work this clamp down the rubber hose toward the throttle body, and remove the rubber hose from the hard line at the heads. There shouldn't be any pressure in the cooling system; very little, if any, coolant should run out.

Take the throttle body with the attached rubber hose over to your workbench.

Use pliers to compress the spring clamp holding the rubber hose to the throttle body's hard line. Remove the rubber hose from the throttle body and install it on the new throttle body (or your freshly cleaned one).

Take another minute to inspect the clean, or new, throttle body. Make sure nothing obstructs the throttle blade and that the throttle plate moves freely.

Make sure the mounting face for the throttle body gasket is clean and totally smooth.

Place the throttle body in the engine compartment and slide the rubber hose back onto the hard line from the heads. This is tricky and may get frustrating; just make sure you seat the hose all the way up the hard line. Use the pliers to compress the spring clamp and slide this clamp back up the hose. Make sure it's actually clamping the hose to the hard line.

Replace the three bolts that hold the throttle body to the intake manifold. Get each bolt started before tightening it down to 9 ft-lb. Remember, the throttle body is aluminum, but the intake manifold that you are bolting it to is composite plastic. You can easily break the intake plenum if you over tighten the bolts.

the throttle body forward and let it rest. Place the bolts on the windshield tray so you can find them later.

Now comes the hard part. Use the needle-nose pliers to compress the gray spring clamp that holds the short rubber hose from the coolant lines at the heads to the throttle body.

This is the underside of the throttle body, showing where the coolant hoses attach. The hot coolant keeps the throttle body from icing up in cold winter climates.

Reconnect the PCV line to its fitting on the throttle body.

Now carefully slip the return hose to the radiator back onto its fitting. Make certain that this hose is seated all the way up the hard line, and use the pliers to reseat the spring clamp properly.

Reconnect the TPS sensor connector on the passenger side. Make sure that the clip snaps into place properly and is locked in place.

Reconnect the throttle wire connector to the electric motor on the driver's side. Again, make sure that the clip snaps into place.

Replace the coolant overflow tank cap.

Replace the air intake bridge and check that it's seated firmly.

Reconnect the red positive battery terminal.

Now relax and look at all the coolant lines, electrical connectors, rubber hoses, and bolts. Take your time and make certain that everything is back in place and everything is firmly fastened together.

Start the car and inspect the coolant lines for any leaks. If there's leaking, shut the car off and correct the problem.

Check the coolant level and add coolant/water if necessary.

Drive the car around the block or up and down the street a few times, just to make sure no codes have been thrown with the addition of the new or cleaned throttle body.

Coolant Bypass

You can get a minor increase in performance by eliminating the coolant circuit that goes to the throttle body. This system was designed to keep the throttle body from icing up in cold weather.

All of the big mail-order companies carry a kit that you can use for this modification. Make sure the little tube has barbed ends. They'll keep the hose in place a little better than the other types.

Once you look at how all this goes together, you'll quickly see how the little metal tube acts as a bypass for the passage that runs through the throttle body. No longer will the hot coolant circulate through the throttle body. This, in turn, means that the air entering the intake plenum will be slightly cooler. Cooler intake air should result in a very minor horsepower increase.

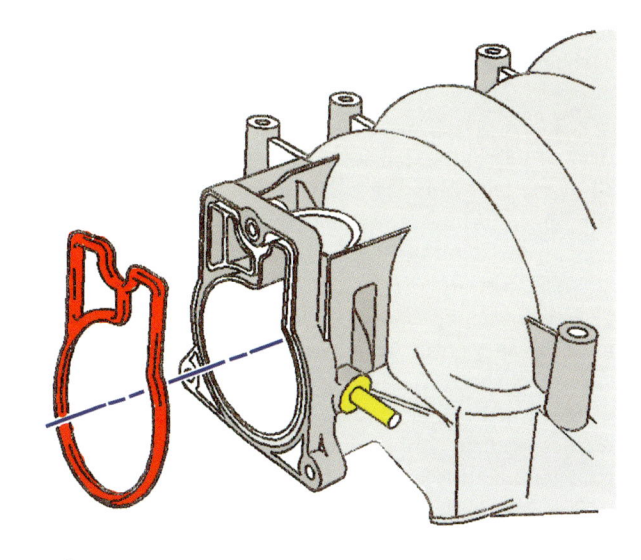

Replace this gasket when you return the throttle body to the intake manifold. *Courtesy of General Motors*

PROJECT 8 | Replacing the Thermostat

Time: 1 hour

Tools: Metric socket wrenches

Torque: Water pump bolts: 10 N-m (89 in-lb)

Talent: ★★   **Applicable Years:** 1997–2004   **Tab:** $60

Tip: Use a rubber or rawhide mallet to tap the thermostat housing if it's reluctant to come off. Don't use a steel hammer.

Performance Gain: None

Complementary Work: Replace the upper radiator hose at the same time you replace the thermostat. If the engine overheated because of a stuck thermostat, change the engine oil, as it may have reached critical temperatures during the overheating.

More confusing things have been said about thermostats than any other part of the C5 Corvette. Let's take a minute to dispel one major myth: The thermostat in your Corvette does not set the coolant temperature. You only need to change the thermostat if it's stuck in either a closed or open position. Your Corvette produces heat energy every time combustion takes place. If we left all that heat in place, the entire engine would melt down in a very short time.

In an effort to save the engine from melting down, combustion chamber heat is transferred to the coolant that circulates around the engine. But eventually, this coolant could become too hot. After all, the coolant can absorb only so much heat. To prevent that, all of the hot coolant from the engine passes through a series of radiator tubes that have little fins on them. The heat from the coolant is transferred to the fins and then dispersed into the atmosphere.

Your Corvette's cooling system is really nothing more than a study in the transfer of heat energy. We need to move all of the heat from the combustion chamber out into the atmosphere before the engine melts down, and that's what the radiator and hoses are all about.

The thermostat is nothing more than a gate in your Corvette's cooling system. When you start your car every morning, the coolant that's resting in the engine block is the same temperature as the air in your garage. The thermostat is designed to hold all the water in the engine block until it reaches a certain temperature. Once the coolant reaches that temperature, the thermostat is opened and coolant is allowed to flow around the entire system, including the radiator.

The stock thermostat opens when the water reaches 192 degrees Fahrenheit. The so-called performance thermostats open the gate at 178 degrees. The normal operating temperature for a C5 Corvette is around 210 degrees. This means that the thermostat has no influence on the temperature of your engine's coolant after things are at operating temperature.

The only ways to lower the temperature of your Corvette's cooling system are to get more air to the fins on the radiator or to increase the amount of surface on the radiator. Changing the thermostat will have no effect on the temperature of the coolant once it's in a fully open position.

The operating temperature of your engine is determined by the size of your radiator and whether or not you have an

More myths have been promulgated about thermostats than any other part on the Corvette. There isn't any real performance gain from installing a thermostat.

The radiator hoses are very easy to change. You can also reuse the factory hose clamps. There's no need to revert to the old-style screw clamps.

oil cooler. It's also a function of how much air is passing through the radiator. Moving more air over that radiator will result in cooler temperatures.

If you install a 180-degree or a 160-degree thermostat, all you're doing is opening the gate a little sooner. In the winter this means your coolant may never get warm enough to allow the oil to function at optimum temperature. It also means that your heater won't be operating at peak efficiency.

There is an optimum temperature for your engine. Running the coolant as cold as you can get it makes no more sense than it does to run it extremely hot all the time. Temperature is especially critical to the engine oil. It has to be hot enough to get rid of any accumulation of moisture.

The C5 thermostat is a part of the water pump intake pipe. When you replace the thermostat, you're really replacing the water pump intake. GM doesn't recommend replacing the thermostat on the 1997–2003 cars. Rather, GM recommends that the entire water intake assembly be replaced. This isn't a big problem, since the aftermarket vendors sell thermostats as a complete assembly—complete with a variety of temperature offerings. The 2004 Corvettes and current models have a separate thermostat at half the price of the earlier versions—go figure.

The inlet location on all the C5 and C6 Corvettes is designed to eliminate the thermal cycling that was common with the old Corvette small-block design used in the previous generations.

A large spring-loaded bypass valve rests on the bypass circuit seat when the main thermostat is closed. This means that coolant flow is available to the heater core at low engine speeds, which is good for cold climates.

A Simple Modification

The thermostat is really controlled by a wax pellet that expands and acts on the thermostat piston, forcing it open and allowing the water to flow. If you install a spacer between the piston and the wax, the thermostat will open sooner and the coolant will flow a little quicker.

First, drain all of the coolant out of the motor. Then, remove the serpentine belts and the two water pump bolts. The inlet pipe should come right off. If you need help, use a rawhide mallet to tap the inlet pipe. Under no circumstance should you pry the inlet pipe off with a screwdriver or hit it with a big steel hammer.

A new thermostat, complete with the spacer, is available from any of the major mail-order companies. Thermostats run about $70, although the 2004 may be a little more.

Perform this modification at the same time you're doing the throttle body bypass hose to limit the mess of antifreeze. For less than $100, you might pick up a couple of horsepower. Just think of this as a fun way to spend a Saturday in the garage.

This is the thermostat in its housing.

39

PROJECT 9

Reprogramming Your Corvette's Computer PROM

Time: Several hours

Tools: A laptop computer and the LS2 Edit program

Torque: N/A

Talent: ★★★★

Applicable Years: 1997–2004

Tab: $200 to $500

Tip: You really need a dynamometer to assess the changes you make in the programming.

Performance Gain: There's real power to be gained from changing the engine management program. It's not going to be cheap, but it's certainly effective. The secret is finding someone who's had experience doing this before.

Complementary Work: N/A

There are a lot of programmers available for the home shop, and most of them work well. But if you want a real power gain, you need to find a dyno shop and get serious.

There are a lot of things you can do to your Corvette to make it faster, but most of these things detract from the car's general performance. Just because you can alter the parameters in the engine management program, doesn't ensure that your Corvette is going to get faster.

Before you spend a lot of money on an aftermarket computer system programmer like the Hypertech Power Programmer III, remember that General Motors has an incredible amount of highly sophisticated testing equipment. GM also has the largest and most highly skilled engineering staff in the world.

These toy programmers (and they are toys) can make some changes in your Corvette's computer system, but those

The main computer is located up in the right front fender well, directly below the battery. You shouldn't have to access it for reprogramming, since that's all done with a computer.

One of the best ways to check the operation of your C5 is to visit one of the major Corvette shows. There's usually a company that brings a portable dynamometer to the show and spends several days running C5s. Not only is it convenient, but the price is usually very reasonable.

changes won't necessarily constitute improvements. When you connect the Hypertech Power Programmer or any similar device to your vehicle, its program recognizes your vehicle's powertrain by reading the vehicle identification number (VIN). The VIN contains specific information on the engine, transmission, and drivetrain of your C5.

Among other things, the Power Programmer lets you choose a program for regular 87-octane or premium 93-octane gasoline. If you decide to use the 87-octane fuel table, you'll retard the spark advance and lose performance, but you can save money using the cheaper gasoline.

After you select the appropriate fuel table, the programmer automatically installs a program that allows you to reprogram the vehicle's computer to function properly with aftermarket components such as unique tire sizes and rear-end gear ratios not offered by GM.

Fortunately, if you screw anything up you can return to OEM (stock) programming in just minutes. Your biggest challenge is assessing whether what you did actually caused your Corvette to go faster. There are two basic ways to measure Corvette performance. First, you can use a dynamometer—either an engine dyno or a chassis dyno—to measure horsepower. Or, you can run a series of carefully controlled track tests.

Don't assess performance by your reaction as a driver to the changes. The easier a car is to drive the slower it will seem. Check the stopwatches and the computer printouts to see what is really the case.

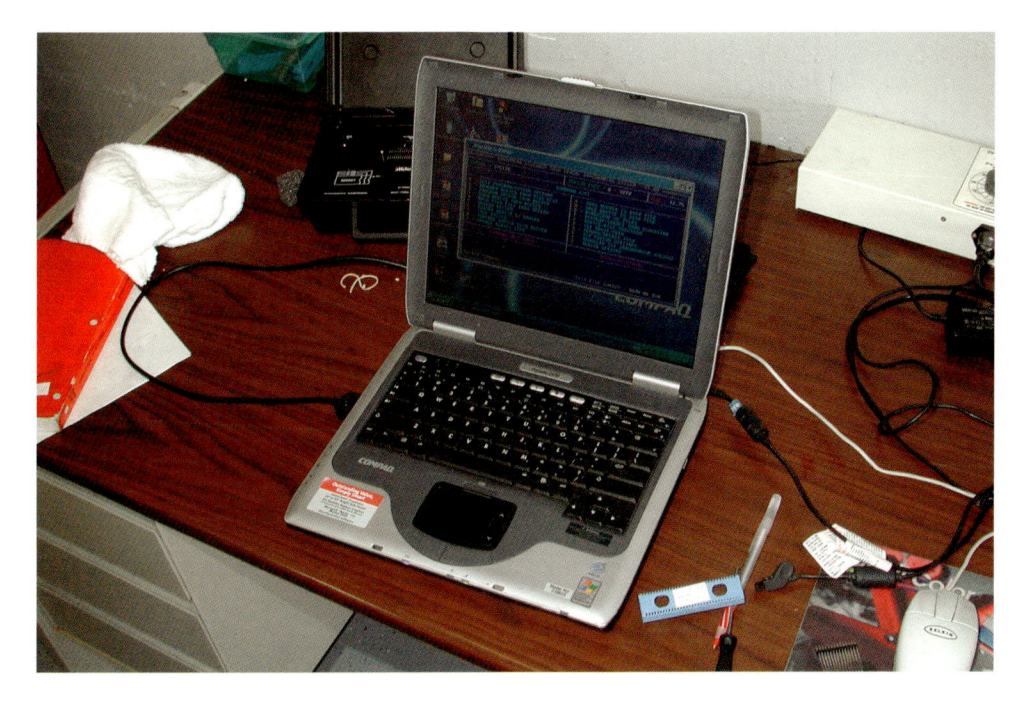

There are three things you need to improve C5 performance—a dynamometer, a computer, and a tuner who knows how to use them effectively. You can only do so much with bolt-on parts. Find someone who knows how to make it all work effectively.

ENGINE

41

PROJECT 10 | Replacing the O_2 Sensors

Time: 1 hour

Tools: Oxygen sensor socket, ⅜-inch ratchet and extensions

Torque: 40 to 44 ft-lb

Talent: ★★ **Applicable Years:** 1997–2004 **Tab:** $40 to $200

Tip: Always remove the oxygen sensor when the exhaust is still hot rather than after the entire system has cooled down. You may need several different socket extensions and a universal joint for your ratchet wrench. Also, make sure you use a proper O_2 socket for the sensor and use anti-seize compound on the threads when you install the new sensor.

Performance Gain: The biggest difference will be in improved idle performance. You could see a performance gain across the board, depending on how badly the O_2 sensor was deteriorated.

Complementary Work: N/A

Your C5 Corvette has four O_2 sensors and each one is fairly expensive. Thankfully, you will probably not have to replace them very often. But an oxygen sensor is cheap enough—relative to the improved performance gain—that you might want to replace the sensors every 50,000 miles as part of a routine maintenance program.

The O_2 sensors create a voltage signal that corresponds to the oxygen level in the exhaust system. The engine management system interprets this signal and uses the signal to keep emissions within the proper range. The tip of the sensor is made of zirconia ceramic and is shielded from the exhaust gases by a tube with holes or slots in it. All the parts in your emission system are critical, but the O_2 sensors may easily be the most critical, since they are the starting points for the engine management system.

Two of your C5's sensors are mounted before the catalytic converter and two more are mounted after the converter. The sensors after the converter monitor the converter activity. This is part of what is called OBD-II, or the second generation of onboard diagnostics.

When replacing a sensor, you may find that it's rusted into place on your exhaust system. This is seldom a problem with a C5, however, since the C5 uses either a stainless or titanium exhaust pipe.

Oxygen sensors need not fail completely but may start to act sluggish. You may notice a slight loss of power, a rougher idle, or a slight drop in fuel mileage coupled with increased emissions.

Almost any shop can read the number of oxygen sensor counts, or the number of times each sensor actually measures your exhaust gases. This foolproof diagnosis might eliminate an unnecessary replacement, although the diagnosis may cost more than a new oxygen sensor.

Before you start to remove an oxygen sensor, remove the negative battery cable. Any time you do electrical work on your Corvette you should disconnect the battery. This is especially true when you start to mess around with the sensors and other such components.

Most oxygen sensors come with the threads already coated with an anti-seize compound. GM uses a special compound composed of graphite and glass beads. The graphite will quickly burn away, but the glass beads will remain. If you have to remove and reinstall the old oxygen sensor(s), make sure that you coat the threads with an anti-seize compound.

O_2 Sensor Simulators

All cars and light trucks made since January 1, 1996, have two O_2 sensors (four in dual-exhaust applications) to measure the amount of oxygen contained in the exhaust gases. The first (primary) sensor is in front of the converter and is used by the onboard computer to calculate the air/fuel mixture. The second (secondary) sensor is downstream from the converter and its only job is to report if the converter is working properly.

A lot of people eliminate the two rear converters, which also eliminates the two rear O_2 sensors, when they install a new H-pipe. They then install simulators to calm down the OBD-II computer system. These simulators are simple electronic devices that fool the Corvette's computer. They plug directly into existing O_2 sensor harnesses.

Unplug the O$_2$ sensors from the wiring harness before you drop the exhaust. Then simply drop the exhaust with the sensors in place. It's a lot easier to remove them with the pipes on the ground.

This is illegal, so be forewarned if your state does emissions testing. If you eliminate the catalytic converters from your Corvette, which violates a federal law, the "service engine soon" light will light up on your dash. You can also use a programmer to disable the rear O$_2$ sensor function from the computer system. That requires more equipment and a little more skill than the simulators require.

The only reason to eliminate these sensors is if you're going to install new H-pipes that eliminate the two rear catalytic converters. It's about as illegal as you can get, but some people feel they really need a trick exhaust system. At least it's better than constantly watching your dash flash the "service engine soon" light.

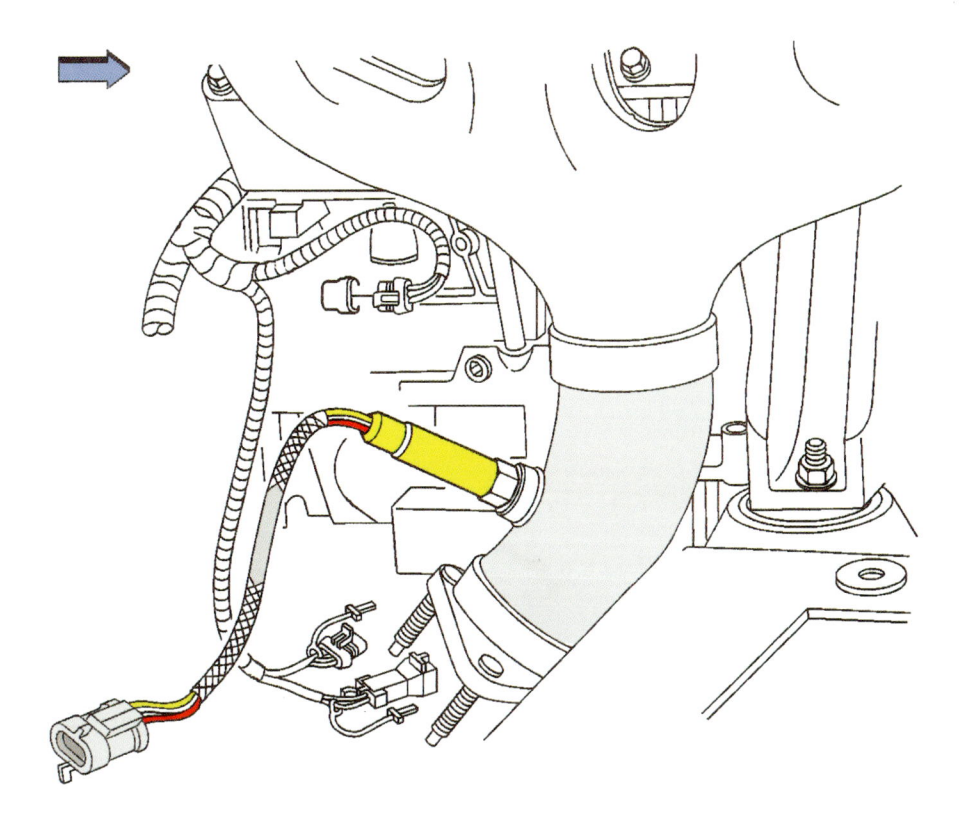

The O$_2$ sensors on the intake manifolds send messages to the main computer, which in turn sets the air/fuel mixture. The sensors in the rear only monitor that the catalytic converters are working properly.
Courtesy of General Motors

PROJECT 11 | Changing the Coolant

Time: 1 hour

Tools: Basic wrenches and sockets

Torque: N/A

Talent: ★★   **Applicable Years:** 1997–2004   **Tab:** $25 to $75

Tip: It's best to stay with Dex-Cool coolant.

Performance Gain: There's no true performance gain, but a well-maintained engine will always operate better.

Complementary Work: Remove and clean the coolant surge tank on the right fender.

Changing the coolant is a cheap and easy reason for you to hide out in your garage for a while. GM says you can go five years between coolant changes, but it doesn't hurt to change it every couple of years.

The LS1 engines at Bowling Green are filled with a mix of 45 percent tap water and 55 percent ACDelco Dex-Cool coolant. Dex-Cool is ethylene-glycol antifreeze with a nonsilicate, anticorrosive chemical package. Dex-Cool was jointly developed by GM and Texaco and became factory fill in most GM cars for the 1996 model year. Its nonsilicate formula is longer lasting and better for durability of cooling system parts than the anticorrosive packages found in old-style antifreezes. Dex-Cool has an orange-pink color.

If only small additions of coolant are necessary, tap water will suffice. If a large addition of coolant is required, find the leak before you do anything else. Then add a 50/50 mix of distilled water and Dex-Cool. I suggest distilled water because tap water in some areas may be somewhat acidic. Distilled water isn't that expensive, so why take a chance?

The C5's coolant change interval is five years, or 100,000 miles, whichever comes first. As good as Dex-Cool is, I suggest changing the coolant in your Corvette at three years or around 45,000 miles. If you change the coolant, mix the Dex-Cool and distilled water at approximately a 50/50 proportion. If you live in a warm climate, you can lower your operating temperature a little by reducing the amount of Dex-Cool in the system. Just don't find yourself in Canada in December with a lower lever of Dex-Cool.

Don't substitute other antifreeze products for Dex-Cool. If you should have a problem and need to use different antifreeze in an emergency, drain the coolant, flush the system with water, and refill with the Dex-Cool/water mix as soon as possible.

Remember that working on the cooling system involves hot liquid. If the engine is at operating temperature, it's possible that when you remove the coolant surge tank cap the loss of pressure will cause explosive boiling in the system that may cause injury. The C5 pressure cap is threaded to allow pressure relief. If the cooling system is hot, turn the cap a quarter-turn. When the hissing stops, turn the cap until it comes off. If the engine is running hot—a coolant temperature of 220 degrees and above—allow the engine to fully cool before removing the cap. Don't take chances with hot coolant.

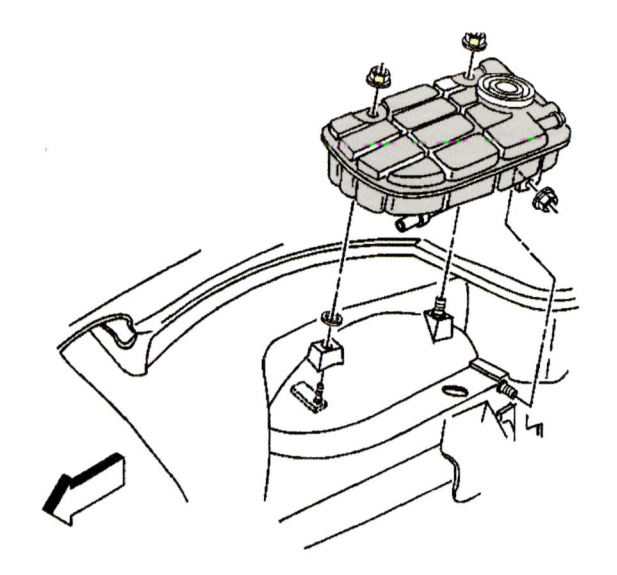

After the coolant is flushed, and before you refill the system, remove the surge tank and carefully clean it with soap and water. These tanks build up an incredible amount of sediment. *Courtesy of General Motors*

GM and Texaco Recommendations about Dex-Cool

Dex-Cool coolant is made by Texaco and bottled under the GM brand. General Motors made a 14-minute training video—*Understanding Radiator Cap and Cooling System Contamination*—to help technicians service common cooling system contamination problems in GM vehicles.

You can buy the video for $10 (plus shipping and handling). Call MSX International of Auburn Hills, Michigan, at 800-393-4831, and ask for part number RADCAPK.

The video offers the following recommendations:

(1) Keep the cooling system filled. In fact, fill the reservoir bottle to the "hot" level when the system is cold. Problems arise when the system's coolant level is not maintained. (Fleet vehicles receiving regular maintenance, and with reservoirs kept slightly above normal, never showed signs of contamination.)
(2) Coolant problems appear to be caused by system contamination, not by the breakdown of Dex-Cool.
(3) Check the pressure cap and keep it clean and functioning. A contaminated and/or malfunctioning cap causes low coolant levels, which in turn cause overheating and a greater loss of coolant. If the cooling system seems to be losing coolant, the first step it to test the pressure cap.
(4) Make sure that the coolant is at a 50/50 mix. Make sure the flush water is removed from the engine block. Any fluid that has been diluted beyond its recommended levels will have a lowered level of inhibitors and won't be able to protect the cooling system effectively. Low levels of inhibitors can cause pitting on aluminum surfaces and general corrosion of cooling system metals.
(5) To achieve a true 50/50 mix, first determine the actual capacity of the system (use the owner's manual). Then add 50 percent of that amount of undiluted Dex-Cool (or other coolant), and top it off with water.
(6) Mixing a "green" coolant with Dex-Cool reduces the change interval to two years or 30,000 miles but will cause no damage to the engine. In order to change back to 100 percent Dex-Cool, however, the cooling system must be thoroughly drained and flushed.
(7) Bacteria cannot live in a hot ethylene-glycol environment and are therefore not a threat to Dex-Cool.
(8) Use a refractometer to check the condition of the Dex-Cool. Its inhibitor package is strong enough that if the batch still provides proper freeze protection, it will probably still provide proper corrosion protection as well.
(9) Dex-Cool can handle the minerals in hard water better than conventional silicate coolants. Potable water is suitable for top-off.

Changing coolant in the LS1 is more time-consuming than it was with older Corvette engines because the engine block coolant drains aren't accessible when the engine is in the car. The left drain is hidden behind the engine mount bracket and the right drain is behind the starter. Because of this, you can't completely drain the system; you can only flush it, so you'll be working with hot engine parts and hot liquid.

The radiator drain is at the lower right corner of the radiator. Open the drain, let it run until the flow stops, and then close the drain plug and refill the system with plain water. Run the engine until the thermostat opens, shut off the engine, carefully release the pressure in the cooling system, and open the radiator drain again.

Repeat the procedure until the coolant coming out of the radiator runs clear. You may need to do this flushing trick several times to get the water to run clear. Once you're done with all the flushing, close the drain and fill the system with a mixture of Dex-Cool and distilled water until it reaches the base of the surge tank neck. You'll probably need about 2 gallons of the mixture.

Now start the engine and let it idle for a minute, and then cycle the engine speed from idle to 3,000 rpm and back until the coolant temperature reaches 210 degrees. Remove the tank cap, observing all the cautions discussed previously. The coolant level should be about ½ inch above the "cold full" mark on the tank. If it's not, top the system off with the Dex-Cool and distilled water mixture. Finally, replace the cap, start the engine, and take the car for a 10-mile drive.

The Surge Tank

The C5 uses a plastic surge tank. Over the years these things get filthy. If you change the coolant in your engine, you may want to remove the tank from the car and thoroughly clean it. You only have to remove the hose and a few bolts. Then take some Simple Green, or another general-purpose cleaner, and rinse the tank out.

PROJECT 12 | Dressing up the Engine Compartment

Time: A few minutes to days

Tools: Most of the aftermarket panels use adhesive backing, so it's simply a matter of sticking them in an appropriate place

Torque: N/A

Talent: ★★

Applicable Years: 1997–2004

Tab: This can vary widely, but spending $1,000 on dress-up items is not unheard of. On the other hand, you can budget $100 a month and make a big improvement.

Tip: Make sure you look at a lot of different Corvettes. The choices are numerous and some of the kits don't fit well. Talk to the people who have already done this job.

Performance Gain: While there's no gain in performance, you sure get a lot of credibility when you open the hood.

Complementary Work: Start this effort with a complete cleaning of the engine compartment. Everyone will notice grease and soil stains before they see all of your new parts.

When it comes to dressing up the engine compartment, you can run wild knowing the car will run the same when you're done as it did when you started.

The variety of options is mind-boggling. Before you begin, decide what you want for the overall appearance under the hood. Carbon fiber and chrome seem to be very popular, but colors are also very popular. Attend a couple of local Corvette shows and find a car with the unique look that appeals to you. Then talk to the owner about where he or she purchased the materials.

Also ask the owner how difficult it is to keep clean. You can generally wipe off faux carbon fiber with a soft towel and some Windex. Chrome panels, however, require a little polishing to keep the shine. The stock engine compartment can be easily cleaned with a little Simple Green or other all-purpose cleaner and a brush. Just keep in mind that when you add a

You can have a tremendous amount of fun with the C5 engine compartment. There are several detail items available, and the possibilities for painting are limitless.

Although faux carbon fiber has become a bit of a cliché, it really looks wonderful in the engine compartment.

This painted air intake duct won't make you go any faster, but it sure looks nice. You can do it yourself if you have the talent, or just order one from the myriad Corvette mail-order parts companies.

dress-up kit under the hood, you're also adding work to your weekly maintenance program.

There's no performance factor to the use of eye-popping panels and covers. Don't choose a dress-up kit that is too hard to keep clean. Onlookers won't be impressed if everything is spotted with water or covered in grease. That's why the first step here is to start with cleaning and detailing. Always use the least-aggressive product possible. Many engine cleaners available in stores are way too aggressive and will damage metal parts, paint, and hoses.

For the same reason, don't bother having the engine steam-cleaned by a commercial engine cleaner. This cleaning process is only necessary as part of a total engine rebuilding process. Your C5 shouldn't need restoration just yet. Even though modern engines are fairly well sealed, high-pressure water can make its way into connectors and electronic components of the engine compartment.

My favorite cleaners are Simple Green and the citrus-based cleaners available in grocery stores. Simple Green may be a little stronger than the citrus cleaners, but it doesn't damage anything. Another favorite of mine is P21S wheel cleaner. This product does a great job on all the cast aluminum at the front of the C5 engine. This product is generally easiest to get online or through a mail-order supplier.

You'll need a good selection of brushes, so head over to the home improvement store. I roam the paintbrush aisle until I have a half dozen or so brushes in a variety of styles and shapes.

Sometimes it doesn't take a great deal to improve your engine compartment. This Corvette's owner did it with just the addition of some red paint. It also helps to keep the engine spotless with an all-purpose or citrus cleaner.

ENGINE

ENGINE

I like to use brake cleaner and a small brush for removing the larger clumps of grease. It's cheap, it's available everywhere, and it won't harm aluminum or plastic. Use the brake cleaner to carefully clean any areas where you plan to attach new dress-up panels to make certain that the adhesive sticks to the panels. Next, rinse the engine compartment with a mild hose spray to remove all the surface dust. Then spray the citrus cleaner over the entire area. I let this sit on the surface for a few minutes before attacking everything with brushes.

Now thoroughly spray down the adjoining painted surfaces (fenders, front fascia, etc.) until all the cleaner and dirt has been washed away. If you notice a few areas that you missed, repeat the process.

Once you're happy with the cleaning job, carefully rinse everything and start the drying process, using either compressed air or a leaf blower. Gently blow the engine dry with a long-tipped air nozzle.

Now dry off everything with a damp towel. You might even want to pop off the engine covers and dry them off if you're not replacing them. Just try to get everything as dry as possible. Next, start the engine and let it run for a couple of minutes.

Once you have the new panels and covers in place, finish the process by lightly spraying the engine and plastic pieces with Stoner Trim Shine or Pledge furniture polish.

Finally, close the hood and wipe down fenders, windshield, etc., with a damp cloth in case there was any overspray. Start the engine and let it run for five minutes or so. Then open the hood and admire your work.

48

PROJECT 13

Replacing the Spark Plugs and Spark Plug Wires

Time: 2 hours

Tools: Anti-seize compound, dielectric grease, and a selection of universal sockets and extensions, plus a spark plug socket

Torque: 7–11 ft-lb

Talent: ★★★

Applicable Years: 1997–2004

Tab: $200 for both the plugs and wires

Tip: Always twist the spark plug retainer to break the seal before you pull it off.

Performance Gain: The idle should be smoother, and low speeds should be more even.

Complementary Work: This is as good a time as any to change both the air and fuel filters.

The C5 Corvette comes with platinum-tip, "100,000-mile" ACDelco plugs. It also comes with 7-millimeter, carbon-resistor-based plug wires. GM says that all of this will last for 100,000 miles, and it will. But if you're looking for a fun little project, spark plugs and wires are really easy to change. Besides, any day in the garage is better than a day at work.

I like NGK spark plugs. They've worked in every car I've tried them in, and you can get them at almost every parts store in America. The NGK TR55 solid-copper-core plugs match the heat range of the stock plugs in the C5 and can be cross-referenced to the stock spark plugs. The NGK TR55VX plugs are the same plug, but they are the platinum-tipped version with a thinner copper core. The platinum coating actually doesn't last long and will break down or burn off if you drive hard. The solid-copper TR55s don't have as much longevity but seem to provide better performance. These solid-copper-core plugs dissipate heat quicker than the

The plastic covers have been removed on this car to eliminate heat buildup. Also note the little socks the owner put on the plug wires to reduce any chance of the plug wires burning. The underhood heat on the C5 is tremendous, and the owner installed long tube headers on the car. Overheating is not a problem with stock cars, but as soon as you start changing things, heat can be a problem.

ENGINE

Spark plug wires are easy to replace on the C5 Corvette. Each wire goes from the spark plug to the coil pack you see here. If you're replacing the plugs, you might as well replace the wires at the same time. It's easy to do and a set of C5 wires isn't all that expensive.

platinum and will deliver a more reliable, more powerful spark. Platinum plugs are intended strictly for longevity.

The NGK plugs come pregapped from the factory, supposedly at 0.050 inch, but I still check them before I install them. The stock plugs are gapped at 0.060 inch. The tips on the NGK plugs unscrew on the top where the plug wire attaches. Make sure these tips are snug with a pair of pliers, but don't get carried away. A moderate squeeze of the pliers and a moderate twist ensures they're not loose.

You can also put a small drop of Loctite 290 Threadlocker Wicking Grade on the end of each plug before you tighten this little cap. It's recommended for electrical contacts and is formulated to seep into tight threads and assembled components. Now let's go step by step through the replacement process.

Step 1: The engine must be cold, making this a nice Saturday-morning job. Explain to your spouse that if you run an errand before removing the plugs, you could damage your Corvette. Cold is best to ensure that none of the aluminum head material comes out with your spark plugs.

Start on the driver's side, since it'll be the most difficult. Remove the plastic fuel rail covers by gently prying them up. They should pop out easily. Pull the large vacuum line out of the brake booster and place it out of the way. Also remove the rubber tube from the "black thing" that is attached to the exhaust manifold. You'll need as much room as possible to remove the hardest plug of all. The number 7 plug, located at the rear on the driver's side, can be a nightmare to remove.

The plug wires are subjected to tremendous heat from the exhaust area and are equipped with metal plug shields. Don't be surprised if you break one (or more) of the wires when you remove them to change the plugs.

Step 2: Remove each of the spark plug wires from the coil pack first. They should slide off with a twist and pull. Once the wire is off the coil pack, reach down and remove the wire from the spark plug. The factory puts a metal heat shield around the wires that will usually come off with the wire. Just twist it about a quarter of a turn and pull at the same time and it should come off easily. Don't worry about the metal shield. The various brands of plug wires use different ways to protect against the exhaust manifold heat. Just keep your old plug wires around in case you ever decide to restore your C5.

Step 3: Once you get all the wires off, it's time to remove the actual spark plugs. Start with the number 7 spark plug. If you have a swivel socket, put it on and use a long extension. Align the swivels as straight as possible, and loosen the plug. Once you have the plug loose, you should be able to turn the swivel socket with your hand to remove the plug.

Step 4: Work your way forward, removing all plugs. Also keep them in order, and take note of their condition. Look to see if they may be running too lean or too rich, etc. Chances are they're going to look perfect. That's just the way things are with modern engines.

You can begin installing new plugs. Once you're ready to install the new plugs (make sure they're gapped correctly), you'll need to put a light coating of anti-seize compound on the threads. The anti-seize compound will keep the plugs from seizing in their threads, making them easier to remove the next time around. On aluminum heads, anti-seize compound is very important. Torque the plugs to 7–11 ft-lb. Or, once the plug is hand tight, turn it another 1/16th of a turn. Just be careful and don't over tighten the plugs.

Never force a plug to thread with a wrench or socket driver. Since the heads are aluminum, you could easily strip out the threads. It's easy to thread the plugs with just a 3/8-inch spark plug socket and a 2- or 3-inch extension. Once you thread them in by hand, you can then use the socket wrench or torque wrench to finish the job.

Step 5: Once your new plugs are in and tightened properly, place the new wires on the plugs and coil. I use a cotton swab to place just a little dielectric silicon grease in both ends of the wires, where the metal contacts are. This will help keep moisture out and stop any arcing in the wire ends. The wires should snap solidly onto the coils and onto the spark plug tips.

Step 6: Once everything is installed, make sure all your tools are clear and make sure you didn't forget something. Go ahead and start the engine and listen for any strange popping or ticking sounds. If you hear a popping or ticking sound that wasn't there before, you either haven't seated a plug tightly enough or you have a wire that's not firmly seated somewhere. Start with the wires and check that they are pushed on firmly. After your engine warms up, eyeball the engine to check that everything is in place and nothing is touching anywhere.

After you drive it for a while, you may smell a slight silicon odor. This is just the silicon grease and the wires breaking in, and it should go away very soon. Hold on to the metal shields in case you need or want to go back to stock spark plugs and spark plug wires.

Never pull the spark plug wire directly off this coil pack. Always twist the connection to break the seal, and then pull it off. The same technique applies to the end of the wire at the spark plug.

PROJECT 14 | Installing an Alternative Intake Manifold

Time: At least a day

Tools: Metric sockets and open-end wrenches

Torque: Intake manifold first pass: 5 N-m (44 in-lb), in sequence; second pass: 10 N-m (89 in-lb), in sequence; valve cover rocker arm cover bolt: 12 N-m (106 in-lb)

Talent: ★★★★   **Applicable Years:** 1997–2004   **Tab:** Over $500

Tip: Make sure that your engine is clean before you begin. You won't want dirt falling into the areas that will be exposed.

Performance Gain: The amount of the horsepower increase will depend on what else you do to enhance the engine's air intake. It's possible you could see a 20-horsepower increase. It's also possible that you see less than 5 if you only swap the manifold.

Complementary Work: Consider installing a larger throttle body on your Corvette. At the very least, install the second-generation Z06 air filter lid.

The stock intake manifold on the early LS1 is a major restriction on horsepower. I've seen increases up to 20 horsepower on the dyno from an intake swap. This shouldn't be surprising, since power gains are really about getting more air into the engine and then getting it out the exhaust as efficiently as possible.

The intake manifold in your Corvette is plastic. Plastic manifolds are easier to mold into complicated shapes, are cheaper to manufacture, weigh less, and run cooler than any aluminum manifold you could find. The aftermarket is taking a while to catch up with this technology.

We call it "plastic," but it's technically Dupont Zytel, Nylon 6.6, which is nylon with glass-fiber reinforcement. It does not transfer engine heat to the incoming air and it isolates the fuel rail and injectors from heat.

The air plenum rests in the space beneath the runners that you see when you open the hood. The stock unit uses 260-millimeter runners that route the air up and back over the plenum into each intake port on the cylinder heads.

The throttle body is attached to the front of the intake manifold and uses a one-piece silicone gasket for sealing. The manifold absolute pressure (MAP) sensor is located at

The different intake manifolds all look the same from the top.

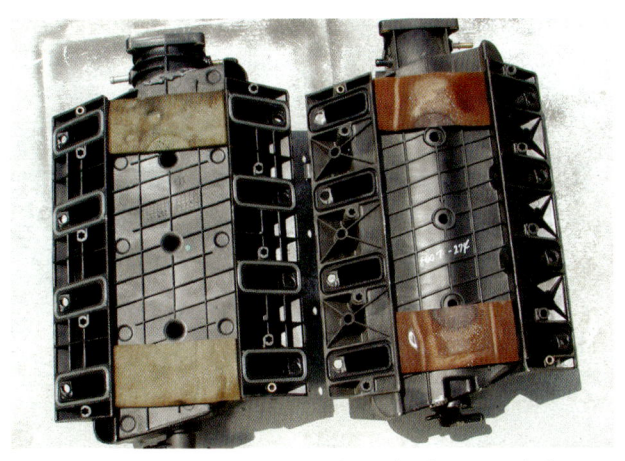

The gain in performance comes from the increased plenum size by pushing the unit lower into the valley of the engine.

The front of the plenum is exactly the same for both intakes.

Use wide masking tape, like that used by body shops, to tape off the rocker arm area when you work on the intake replacement to keep all the dirt out of the cylinder heads.

the rear of the manifold and is sealed by an O-ring. Keep in mind that this LS1/LS6 manifold handles only air; there is no water crossover, and no oil contact due to the sealed valley cover.

The fuel rail assemblies and the eight injectors are sealed in individual manifold bores with O-rings. The injectors are pointed at the back side of the intake valves to promote better atomization.

Before you get excited about replacing this manifold, keep in mind that it's entirely possible to get 400 horsepower using the stock LS1 manifold. You really only need an aftermarket manifold if you want to run very high rpm or you've increased the displacement of the engine.

The LS6 Manifold

Follow me closely on this one—the LS6 manifold is really the LS1 manifold. In 2001, General Motors decided to reduce the number of parts in the engine plant. Since the LS1 and LS6

The opening for the throttle body is the same for both of the intake plenums. Make sure the surface is clean before you put the throttle body back in place. Always use a new gasket whether you're just cleaning the throttle body or you're installing a new one.

ENGINE

53

manifolds cost the same to produce, GM just gave everyone the LS6 manifold in the 2001 Corvette. They were able to reduce the amount of parts in the plant at no additional cost.

The LS6 manifold has greater plenum volume than the LS1 as well as revised runner length. It also eliminates the exhaust gas recirculation (EGR) system. The material and shape are virtually the same as the LS1 manifold, and very few people can tell the difference once the parts are installed.

If you install the LS6 manifold on your early C5, you'll see an increase of 6 or 7 horsepower. The LS1 manifold only becomes a restriction if you've modified the cylinder heads and changed the camshaft. If that's the case, there's about a 20-horsepower difference between the two manifolds.

Some aftermarket companies are making C5 intake manifolds in aluminum. The aluminum, however, picks up the heat from the engine much quicker and transfers it to the incoming air charge. A change to an aluminum manifold is really a step back in time.

If your Corvette already has the LS6 intake system, I would just leave well enough alone. If you get to 400 horsepower at the rear wheels, then there are some different options available for you.

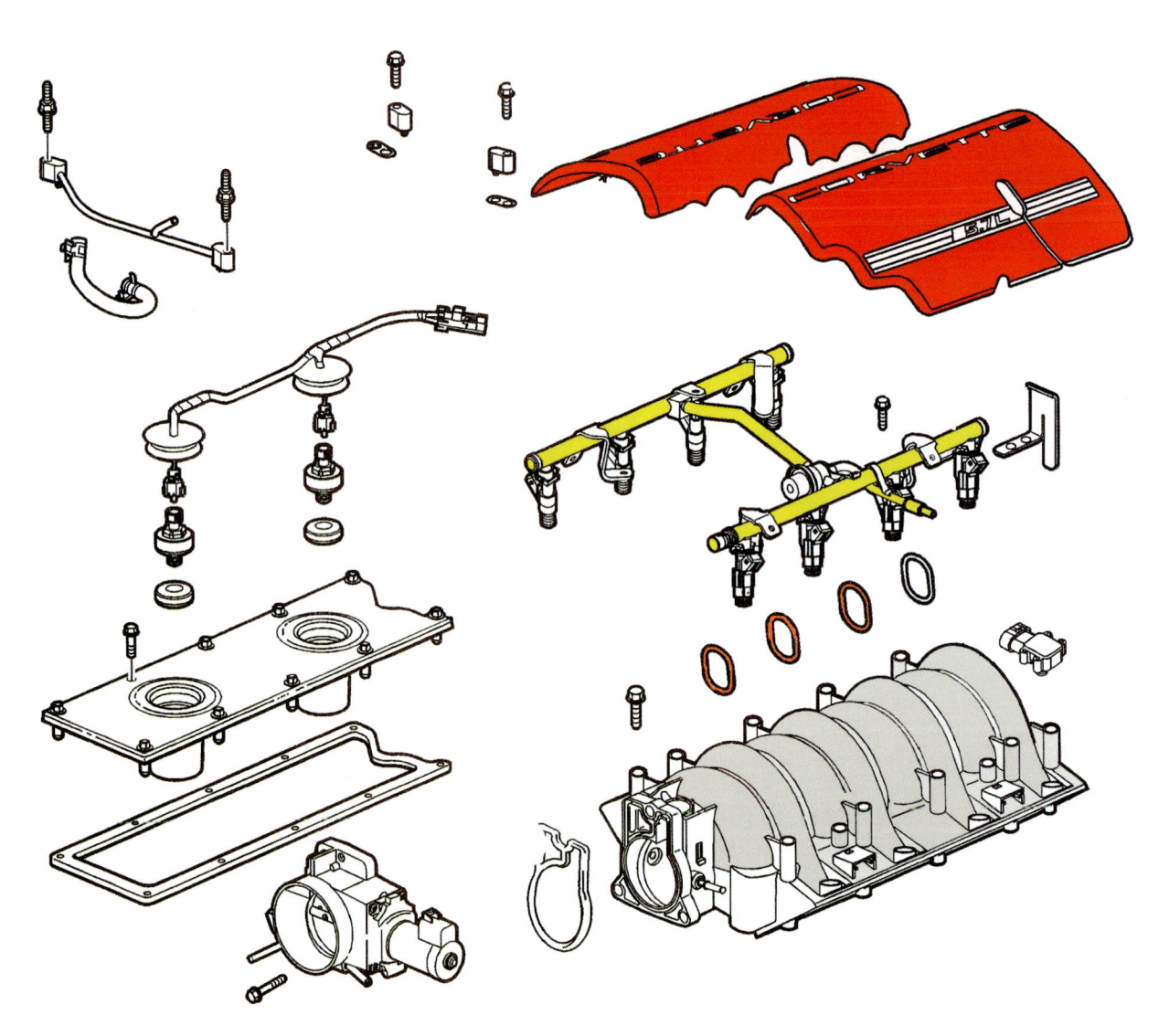

This is how it all goes together. When you look at the intake system this way you see a certain elegance in the simplicity of it all. *Courtesy of General Motors*

PROJECT 15 | Replacing the Fuel Injector

Time: 2 to 4 hours

Tools: Metric sockets

Torque: Fuel rail: 10 N-m (89 in-lb)

Talent: ★★★   **Applicable Years:** 1997–2004

Tab: $500 + for high-performance injectors

Tip: Always lubricate the fuel injector O-rings before installation.

Performance Gain: There will be no real performance gain unless other modifications are made.

Complementary Work: Change the fuel filter on the 1997–2003 cars. (There is no fuel filter on the 2004 model.)

No single topic takes up as much space on Internet forums as injector swapping. The C5 Corvette fuel system has eight injectors, a fuel rail, a 0.375-inch fuel supply line, and an electric pump in the fuel tank. C5 Corvette engines, except for the 1997, which uses a traditional return line, pass all of the fuel down the fuel rail. Fuel pressure in the early engines was regulated to 58.02 psi (four-bar) by a regulator in the fuel tank. The 1997 Corvette was again an exception—fuel pressure was regulated at the fuel rail entrance and referenced by manifold vacuum.

Let's look at some basics: The amount of fuel that an injector delivers is determined by the length of time it's held open and by the pressure of the fuel system. Now consider that an injector must be able to respond quickly to changing demands. It must go from supplying a very small amount of fuel to a large amount at wide-open throttle. And it must do this instantly so as to prevent high rpm lean out, or pistons with holes in them.

The fuel injectors simply push onto the fuel rails and are held in place with little clips. In order to remove the injector, simply spread the clip open.

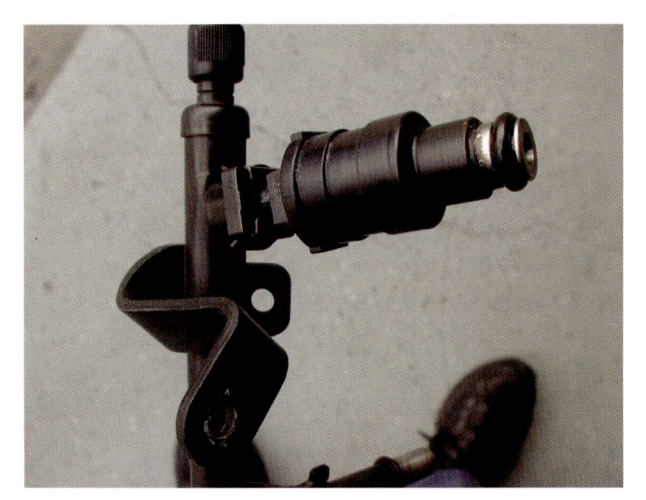

The lower part of the injector, which is pushed into the intake manifold, uses an O-ring for sealing.

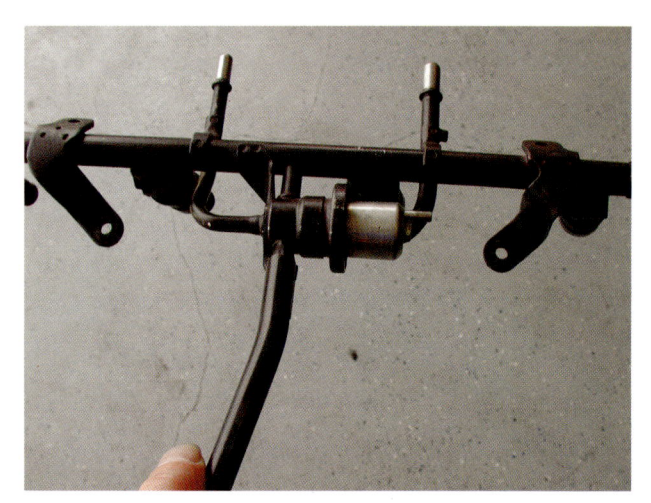

The fuel pressure regulator is part of the fuel rail on the early cars. An adjustable pressure regulator is only useful if you're well over 400 horsepower.

This regulator has been modified. Increased pressure will push more fuel into the combustion chamber; however, this is only true if your engine actually needs more fuel. On a stock engine, the computer will shorten the time the injector is open to compensate for the increased pressure.

Flow Rate

Injectors are designed to flow a certain amount of fuel, measured in pounds per hour. The C5 Corvette uses a variety of different injectors with flow rates varying from 26.4 pounds per hour to 29.1 pounds per hour.

The stock C5 injectors will flow enough fuel to easily make 400 to 425 horsepower. But beware: if you operate an injector at its limit for an extended period of time, such as at track events, it'll wear out faster.

The C5's fuel injectors are easy to swap around, but poorly matched injectors can cause a lot of problems. Do it wrong, and your Corvette won't idle properly and you'll also kill your throttle response. If you intend to swap your injectors around, budget some money for dyno time and a recalibration of the fuel delivery tables in your powertrain control module (PCM).

Now back to some more basics: If you increase the fuel pressure, it will increase the flow of fuel out of the injectors, assuming you haven't changed the PCM command. The PCM controls the quantity of fuel that's supplied to the engine by regulating the amount of time the injector is open. This is called pulse width, and it's measured in milliseconds.

In reality, if you increase the fuel pressure, the computer will shorten the pulse width. You'll end up with the same amount of fuel reaching the combustion chamber as you did at the lower pressure. You only need to increase the pressure if you run at very high rpm for an extended period of time.

You can get a couple of horsepower by adjusting the fuel pressure, but they'll be at the very top of the power curve. You'll never notice it on the street. In modern engines, no single change will transform your car's performance.

Duty Cycle

As the rpm of your engine increases, the window for fuel delivery becomes shorter. For instance, at 4,000 rpm there are 30 milliseconds to deliver all the fuel that's necessary, but at 6,000 rpm there are only 20 milliseconds of time available. At some point, the injector will stay open all the time because it can't cycle fast enough.

The duty cycle is the amount of time that an injector is open. If an injector has a 50 percent duty cycle at a given rpm, then that injector is open half the time. In most cases, you don't want to go much above 80–85 percent duty cycles. The injectors get too hot and become erratic when they have to work this hard.

The injectors on your C5 can produce up to nearly 600 horsepower. You can install all the big injectors you want and it won't do much good for your engine. The only change is that the injectors won't have to work so hard. Instead of being open 50 percent of the time, they may only need to be open 40 percent of the time. Dumping more fuel into a motor

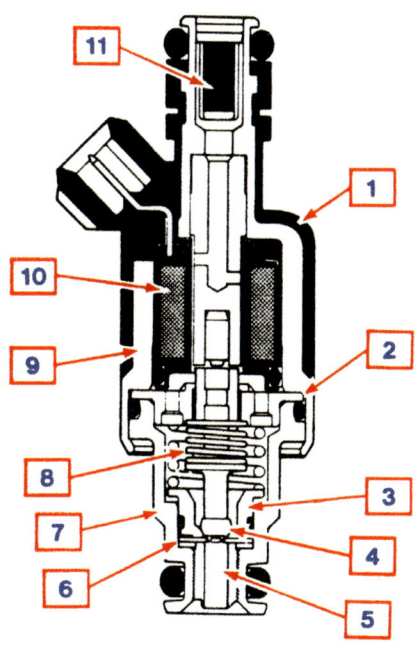

#		#	
1	SOLENOID ASSEMBLY		
2	SPACER & GUIDE ASSEMBLY		
3	CORE SEAT	7	SPRAY HOUSING
4	BALL VALVE	8	CORE SPRING
5	SPRAY TIP	9	SOLENOID HOUSING
6	DIRECTOR PLATE	10	SOLENOID
		11	FUEL INLET FILTER

Injectors are very compact and involve some extremely small passages. It's amazing that we seldom have injector problems. *Courtesy of General Motors*

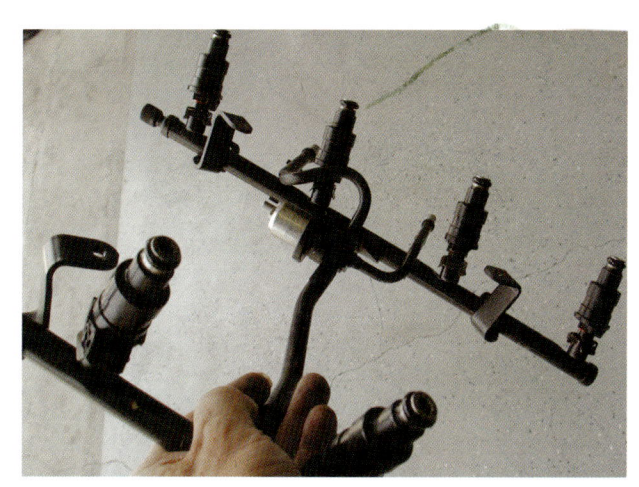

Remove the entire rail, and all eight injectors, as an assembly. Just do it slowly and carefully.

The part number is molded directly into the side of the fuel injector.

won't make more power unless you also increase the amount of air that goes into the motor.

If you're determined to spend some money, consider purchasing a set of injectors that have been flow-tested and matched to within a tolerance of 1 percent. Most injectors on the market are built to within a 5 percent variation. The matched set will give you a bit of improvement but not enough that you'll notice.

For really serious horsepower, the secret is to call Lingenfelter Performance Engineering and get a set of the American Speed Association (ASA) injectors. They were developed by GM for ASA racing, and they work. They're 400-kilopascal (kPa) constant-pressure Bosch EV6 injectors. But even these injectors need a new camshaft, some PCM programming, and sufficient dyno time to get serious horsepower.

Injector Types

Pintle: The pintle is the most common type of injector used by the major manufacturers. A small needle, or pintle, moves in and out of a matching orifice. When the injector solenoid is energized, the needle is pulled back, allowing fuel to spray out. You end up with a moderate spray pattern that works in almost every type of engine.

Disc: Both Bosch and Lucas favor the disc design. They use a little disc that covers another flat plate with a series of holes drilled in it. The actuation is similar to the pintle-type injector. The Lucas injector moves the disc into the body of the injector to provide for a quicker response. Disc injectors generally use a very narrow spray pattern and work nicely in the LS1 engines.

Ball and socket: Rochester, a division of GM, makes a ball and socket injector that provides for better atomization and a wider spray pattern. It is, however, more expensive.

I believe that 99 percent of the injectors ever installed in a C5 Corvette engine are still working nicely. In 130,000 miles, I've replaced only one injector on my car. Fortunately, when the time comes, C5 injectors are easy to replace. If you feel the need to make a change, find a shop with a dynamometer and the ability to edit your Corvette's computer, and a shop owner with the expertise to do it right.

PROJECT 16 | Installing a Louder Exhaust System

Time: 1 hour

Tools: 3/8-inch drive metric sockets

Talent: ★★ **Applicable Years:** 1997–2004 **Tab:** $500

Torque: Oxygen sensor: 42 N-m (30 ft-lb); exhaust manifold to exhaust pipe: 20 N-m (15 ft-lb); exhaust pipe hanger and brace bolts: 50 N-m (37 ft-lb); muffler hanger nuts: 16 N-m (12 ft-lb)

Tip: Take a long look at the rear exhaust tips before you tighten everything. It would be wrong to have the tips off center.

Performance Gain: Nothing you'll be able to feel on the street, but the noise can give you the illusion of a whole lot more power.

Complementary Work: Make sure you have a clean air filter.

A new muffler may be the single most popular modification made to the C5 Corvette. Only wheel and tire changes would compete for popularity. And it's not only cheaper to change your muffler than your wheels, it's actually easier.

Before you purchase a muffler for your C5, I strongly advise that you attend several Corvette shows and talk to the car owners. Ask them how they like their mufflers. Ask them how they enjoy them on long trips, and then ask the same question of the spouses. You might get different answers.

You'll be hard-pressed to feel any horsepower gains from this modification. C5 muffler swaps are all about noise.

The aftermarket exhaust systems have traditionally consisted of a rear muffler section that bolts up to the rear H-pipe. This means that your new mufflers will be in the same position as the stock mufflers. This transverse position creates a small (very small) increase in power and very little increase in interior noise.

Most of the aftermarket exhaust systems are offered in stainless steel, and in a polished form. If looking good is important to you, the polished stainless mufflers look wonderful, particularly the Corsa and Callaway units.

A few manufacturers (Corsa and Borla) are offering a system where everything tucks up into the driveshaft tunnel. Both Borla and Corsa use titanium in these systems. It's interesting to note that when GM puts an aftermarket system on any of its Corvettes, the systems are always from Corsa. I've driven a Z06 Corvette with the central system constructed from titanium and found the interior reasonably quiet.

The H-Pipe

The stock Corvette exhaust system uses an H-pipe system between the catalytic converters and the muffler system. This is another

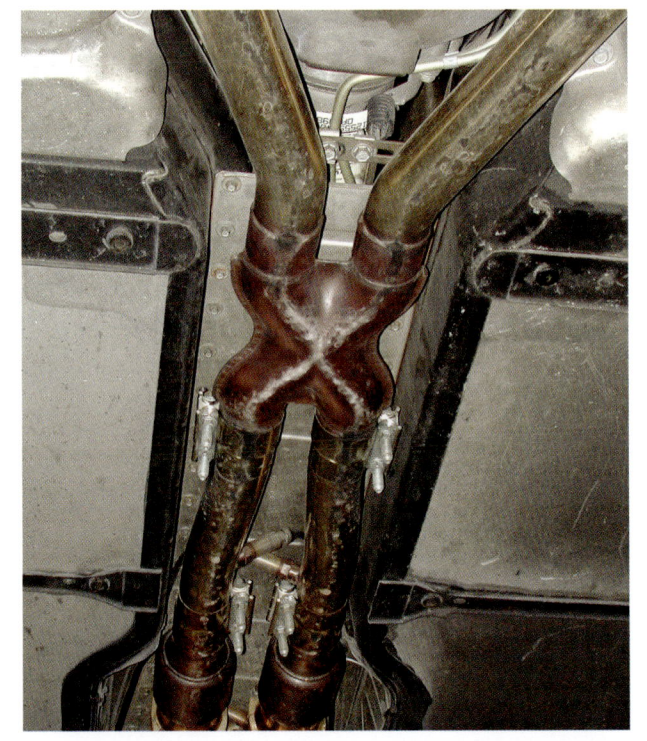

This is the legendary X-pipe that was first developed by Dr. Gas. Notice how it's curved and allows for a smooth flow of exhaust gases. The first person to drive a car with this system installed was Sterling Marlin, when he won the 1995 Daytona 500. No one fully understood how the system worked until Bobby Labonte went upside down in the 125-mile qualifying race that year. Pretty soon all of the NASCAR garages were linking the left bank of cylinders to the right bank with a curved cross-over pipe.

You should be able to swap mufflers on your C5 Corvette in half a day, right in your driveway. The professionals can do them in less than 30 minutes.

The Callaway exhaust system uses unique D-style tips on the mufflers. Tip selection is as important to some people as the sound is.

ENGINE

place where modifications can mean real power gains. Dr. Gas first used this idea back in 1995 when Sterling Marlin won the Daytona 500. When they first tried the system, it improved the power by 4.5 horsepower. In the world of NASCAR, that sort of gain from a single component is almost unheard of.

The Corvette engine, like all the Nextel Cup cars, has a 90-degree crankshaft. There are four connecting rod journals on a V-8 crank, and they're all 90 degrees from one another. This makes a balanced engine, which is important for a passenger car. On each bank there are two cylinders that fire within 90 degrees of each other.

These two exhaust pulses are very close together as they travel through the exhaust pipe. Some V-8s with dual exhaust pipes even make a popping noise when the two pulses arrive at the end of the pipe at roughly the same time. If the two pulses arrive at exactly the same time, though, it can cost horsepower. The effectiveness of the exhaust system is all about the timing of the pulses.

You'll often see a cross-over pipe on V-8s with dual exhaust. This pipe connects the exhaust pipes from the two banks of cylinders. A cross-over pipe smoothes gas flow and lowers sound levels. Dr. Gas (www.drgas.com) figured out

ENGINE

This is a stock system on a 2004 Z06 Corvette. Notice how GM tucked everything up and out of the way.

Drop the rear sway bar out of the way when installing new mufflers. Remove two nuts and two bolts, and the sway bar simply drops out of the way. This might be a good time to lubricate the rear sway bar bushings as well.

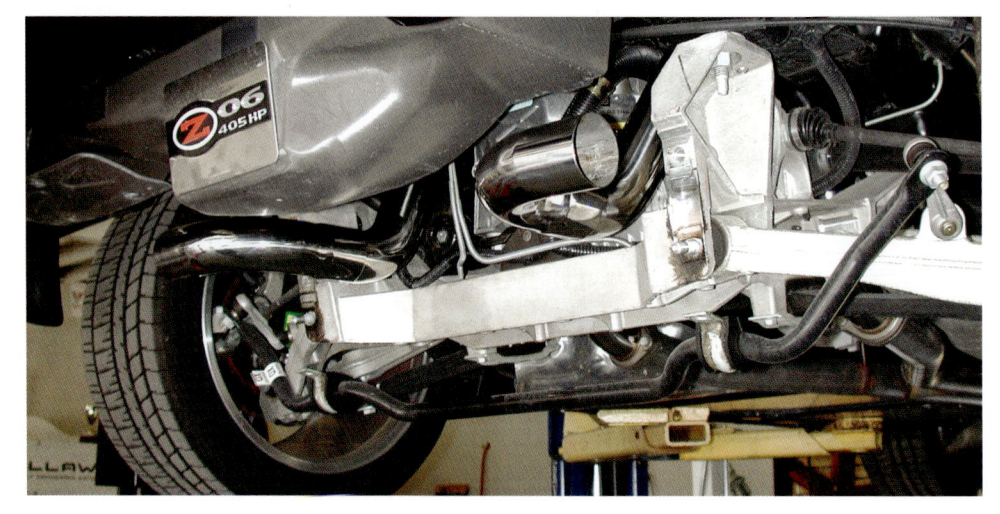

how to use this cross-over pipe to make some extra power. He calculated that a well-designed single pipe increases the overall volumetric efficiency of the engine by about a half a percent—that's a lot of power for a bolt-on product.

A Dr. Gas X-pipe is not the same as an H-pipe, and you may need to take your Corvette to a professional shop to have the pipe welded in place. Considering that this part costs around $200 and will produce as much power as a $1,000 muffler system, this would appear to be a wise purchase.

The Bassani (www.bassani.com) full-length X-pipe exhaust is a very close copy of the Dr. Gas system and is nearly as good. The Bassani system is pretty effective at scavenging exhaust gases, but consider yourself warned: The system violates federal laws regarding emissions. The system is intended for off-road use only, but that hasn't stopped people from putting them on their street Corvettes.

60

Mid America Motorworks installs hundreds of new C5 mufflers at their annual Funfest Corvette party. These guys can put a new muffler on in less than 30 minutes.

Exhaust tips can make a tremendous difference in appearance. Aftermarket manufacturers couldn't keep up with demand the first few years of C5 production.

ENGINE

The C5 exhaust can be easily removed with two people. Once the mufflers and hop-over pipes are removed, simply unhook the O_2 sensors and exhaust manifold bolts. Then the entire assembly drops down.

The stock muffler and the hop-over pipe come off as one unit. Most of the aftermarket systems consist of three separate pieces. They require a little more coordination during installation, and it's good to have a couple extra people around if you do this project at home.

PROJECT 17 | Installing Headers

Time: Several hours to several days

Tools: ⅜-inch drive metric sockets

Torque: Exhaust manifold bolts: 25 N-m (11 ft-lb); oxygen sensor: 41 N-m (30 ft-lb)

Talent: ★★★★   **Applicable Years:** 1997–2004   **Tab:** $60 to $1,000

Tip: If you're investing this much time and effort, you may as well install a complete exhaust system.

Performance Gain: You can gain at least 15 horsepower with long tube headers and different mufflers.

Complementary Work: This is as good a time as any to replace the muffler and add an X-pipe from Bassani or Dr. Gas.

Exhaust headers will make a slight difference in the performance of your C5 Corvette. This is one of the most common modifications made to the C5 Corvette and makes a lot of sense if you do it right.

Think of your Corvette engine as one huge air pump. You take air in and then you move it out. Basically the more air you can take in, and the faster you can get it out of the engine, the better your Corvette will run. That means that once you have a decent air intake system you should move to the exhaust system. Save the internal modifications until later.

Exhaust headers are considered part of the emissions system, and depending on where you live, they may not be legal. The catalytic converters, and anything in front of them, are controlled by emissions regulations.

Don't waste your time and money installing the headers called "shorty" headers. About the only thing they're good for is impressing people when you open your hood. You'll spend a sizable amount of money and get very little performance gain.

If you decide to install headers, make sure that you go for the long tube systems. They're not the easiest items to

This Callaway shorty header system is a piece of art, but this installation won't produce a lot of horsepower.

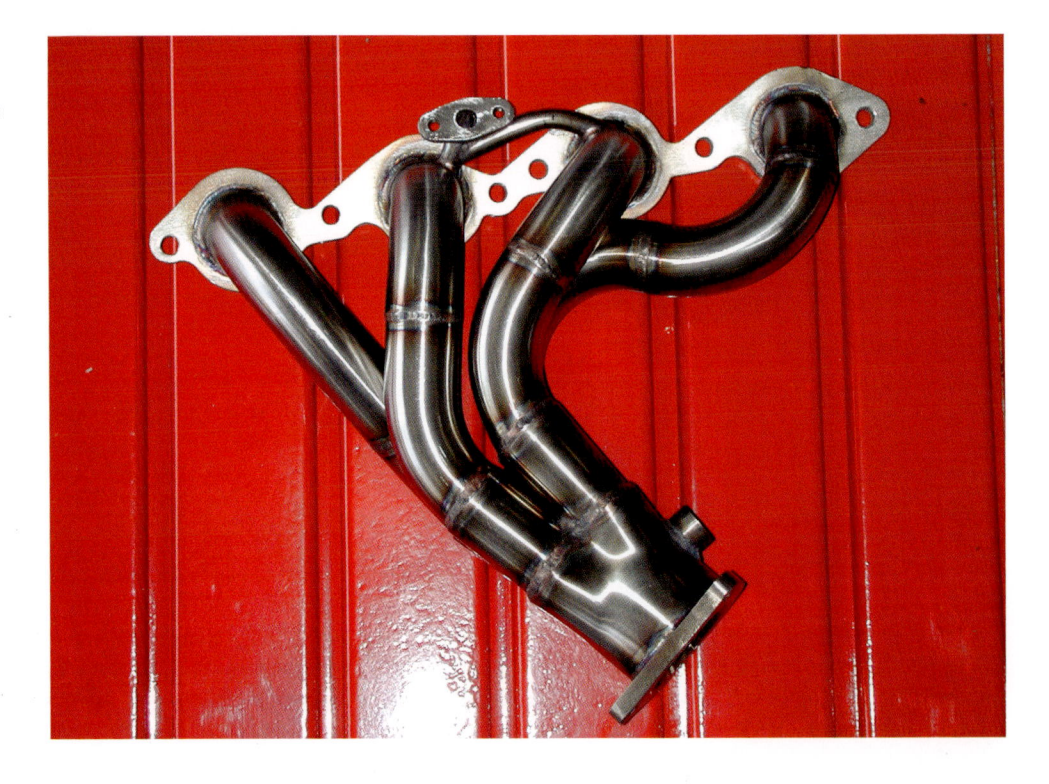

While long tube headers cost a lot more than the shorty headers, they produce a lot more power. Just be careful of emissions regulations.

install, and they're expensive, but they make the effort worthwhile. There isn't much difference in the power gain between the various manufacturers. The most popular headers are produced by TPIS (www.TPIS.com). I've been buying stuff from TPIS for years, and their products always seem to work.

You're going to need a lift for this job, or at the very least some very tall jack stands. The headers have to be angled down into the engine bay, which makes it all a little tricky.

The first thing you'll need to do is disconnect the battery, because you have to disconnect the oxygen sensors. The front O2 connectors are located above the exhaust pipes just to the rear of where they connect to the exhaust manifolds. Remove the blue retaining pin first and then unplug the sensor.

Next, remove the three nuts that hold the exhaust pipe flange to the exhaust manifold. There are three nuts per exhaust pipe that need to be removed with a 15-millimeter socket. You should leave one nut on each pipe while you disconnect the rest of the exhaust system. Once all the bolts have been removed, and the rear of the exhaust is laid on the floor, you can easily undo the two loose nuts and lower the rest of the exhaust.

Remove the exhaust bolts at the other end of the exhaust pipes. There are two bolts per exhaust pipe that need to be removed with a 15-millimeter socket. They are tightened to 50 N-m or 37 ft-lb. Next, locate the forward exhaust hanger just behind the clutch bell housing. There are two 15-millimeter bolts that need to be removed. Next, remove the two rear exhaust hanger bolts with a 13-millimeter socket. Undo the two loose nuts from the front of the exhaust assembly and lay the exhaust on the floor. Slide the exhaust assembly out from under the C5 and put it out of the way.

Place a jack under the front of the car and jack it up enough to remove the front jack stands. Let the front tires sit on blocks or the ground. This allows for easier access to the engine compartment.

Go to the passenger's side and remove the oil filler cap. Then pull up on the plastic cover to get access to the engine. Do the same on the driver's side, but be careful as you bend the cover to not damage the fuel line(s).

Next, identify the air valve on the driver's side. Insert a screwdriver into the plastic clam in order to free it, and remove the rubber tube from the valve. Use a 10-millimeter socket to remove the air valve from the air connector on the manifold. Set this air valve aside.

On the passenger's side, identify the air connection. It's a metal tube that's difficult to bend and move out of the way. Remove it from the manifold using a 10-millimeter socket. You'll need it out of the way later on when installing the headers.

Remove the dip stick–mounting bolt with a 10-millimeter socket and pull the dip stick up out of the oil pan with a firm tug. Next, remove the spark plug wires from the coil packs. Pull on the rubber boot, and not the wire, as the wire can separate from the boot, requiring a replacement. Remove the spark plug wires from the spark plugs and set them aside. Again, pull on the boot and not the wire. Twisting while pulling makes it easier.

Now disconnect the white harness connector and pull the harness retaining clips from the coil pack mounting bolts. Use a 10-millimeter deep socket to remove the coil pack mounting bolts. Early C5s have two bolts per coil pack; later models have a mounting bracket holding them together, and removing the bracket is all that's required.

Go to the driver's side and remove the plug wires from the coil packs and the spark plugs. Again, don't pull on the wires, but use the rubber boots. It's easier to remove the plug wire on the left by moving the alternator out of the way.

Remove the belt from the alternator on the driver's side by disconnecting the plug at the top/back of the alternator, and use a 10-millimeter wrench to disconnect the hot wire from the alternator. Next, use a 13-millimeter socket and remove two alternator mounting bolts from the back of the alternator.

Use the same socket to remove the two long alternator mounting bolts from the front of the alternator. Use the socket to remove the alternator mounting bracket from the engine. Insert a pry bar or screwdriver between the bracket and the alternator to free it. With the alternator out of the way you have access to the plug wire on the left-most coil pack. Remove it and set it aside. Disconnect the white coil pack harness. Remove the wiring harness retaining connectors from the coil pack mounting bolts.

Use a 10-millimeter socket to remove the coil packs from the valve covers and set them aside. A 10-millimeter wrench will also be needed to disconnect the coil pack bolt under the fuel line(s).

Get under the car and disconnect the O_2 sensor from the manifold on the driver's side. It's mounted to the frame with a plastic retainer. Pull it from the frame. Do the same with the passenger's side connector. Use a 15-millimeter socket to remove the steering connector. Swing the steering shaft toward the driver's side of the car in order to make room for the manifold removal and subsequent header installation. Use a 10-millimeter socket to remove the six bolts from the manifold. The rear-most bolt is harder to remove, due to the fuel lines and vacuum hoses that are in the way, so remove it first. Leave a middle bolt in place and remove it last so that the manifold is balanced and requires less effort to hold. Pull the manifold up and away. It will require some maneuvering to do this. Hold the fuel lines away so that you don't damage their metal braid.

Perform the same steps on the passenger's side to unbolt and extract the manifold on that side. The air tube may get in the way, so pull it clear of the manifold. It doesn't seem to want to break easily, but pull on it carefully.

With everything on the garage floor, remove the O_2 sensors from each manifold. Penetrating oil helps free them from the manifolds.

Now you can return to the passenger's side, as this should be easier to install since there's less stuff in the way. It's necessary to remove the valve covers on some C5s in order to install the headers. Use a 10-millimeter socket to loosen the valve cover bolts. Early C5s have perimeter bolts, while the newer C5s have center bolts. These center bolts will stay in the cover with a retainer. There are two rubber tubes that need to be removed from the valve cover. Twist and pull. Set the cover aside and clean up the gasket if necessary.

This little device is called a simulator. When you illegally remove the catalytic converters, the Corvette computer will set off a dash light. The federal government had the automakers develop the OBD-II emissions system to alert the driver of an emissions failure. When you strip the cats off your new Corvette, you can simply plug the simulator into the wiring harness where the O_2 sensor once resided. You won't have to stare at the "service engine soon" light, but remember, this modification violates federal emissions law.

Do the same on the driver's side, but be sure to remove the connector and retaining piece for the wiring harness. Otherwise, it'll be hard to remove the valve cover. There is a rubber tube connected to the right side of the valve cover that will need to be removed. It should come out if you twist and pull.

Place a towel or rag over the head to prevent the header from coming into contact with the rocker arms. Now work the header down into the opening. Have someone on the ground to guide the end of the header and inform you of what's happening down there. All of this will require some pushing and wiggling and you'll probably scratch the headers. Because of all the movement, install the headers before installing the O_2 sensors. You don't want to break them off. Don't bolt the header to anything, though, until you install the O_2 sensors.

Have the person under the car remove the oil filter to allow for the driver's side header to be installed. If you don't do this, the mounting flange on the header won't be able to clear the filter. There will be some oil in the lines above the filter, so use a pan to catch the fluid. (The 6 quarts in the oil pan will not come out when you remove the filter.) Work the driver's side header down just like the passenger's side, holding the fuel lines out of the way. Then install the O_2 sensors into the headers.

Use the manifold gasket that came off with the headers and use the six original bolts to attach the headers to the head. Do not tighten this down yet.

This is the ultimate in a C5 exhaust system, installed in the Pratt and Miller C5-R Corvette.

Do the same on the passenger's side. Reinstall the driver and passenger side valve covers using a 10-millimeter socket. Be sure that the gasket stays in the channel on the valve cover. While on the driver's side, tighten the six header bolts to 18 ft-lb. Also, reconnect the steering linkage and tighten the bolt to 25 ft-lb. On the passenger's side, tighten the header bolts to 18 ft-lb.

Identify the hole in the block and insert the oil dip stick. It should snap into place. Install the retaining bolt through the dip stick mount and into the header, so that the dip stick is secure. Install the coil packs with the 10-millimeter bolts, and connect the wire harness white plug and the wire harness retaining connectors.

Attach the spark plug wires to the plugs and then connect them to the coil packs. Attach the air fitting to the header using the original 10-millimeter bolts. Install the air valve to the rubber hose. Install the plastic retaining ring around the hose. Attach the valve to the header using the original 10-millimeter bolts.

The long front 15-millimeter bolts can now be installed into the alternator to hold it in place. Install the rear alternator bracket to the head and then to the back of the alternator. Now tighten all of the bolts that hold the alternator and bracket in place.

Slip the serpentine belt over the alternator, making sure that it follows the original path over the other pulleys. Once over the other pulleys, the belt can be draped over the tensioner bolt while you get your 15-millimeter socket. Using a 15-millimeter socket, turn the tensioner toward the driver's side and pull the belt over the tensioner. Double-check that the belt is properly mounted over the pulleys before continuing.

Jack the front of the car back up so that you can tackle the undercar work. Tighten the bolts that hold the headers to the bell housing. Between the headers and the frame, reconnect the O_2 sensor cables.

Catalytic Converter

A lot of people use this opportunity to install new Random Technology (www.randomtechnology.com) catalytic converters. They should simply slide into the header. Just make sure that the O_2 sensor fitting is oriented in the same position as the original exhaust. Slip the X-pipe onto the catalytic converter assembly. You may have to pry the catalytic converters away from the frame since they may not be close enough for the X-pipe. The X-pipe should slip on easily at first but then require a little hammering at the ends with a rubber mallet to get it to slide all the way on.

When possible, reconnect the retaining bolts to the X-pipe. If the pipe isn't forward enough, the springs on the hangers help pull the exhaust forward. It may be necessary to wire the muffler connectors away from the X-pipe in order to get the pipe in.

Continue to hammer on the rear of the pipe with a rubber mallet while wiggling the assembly until the X-pipe is firmly attached to the catalytic converter assembly. Next, remove the two O_2 sensors from the original exhaust and connect them to the new exhaust, routing the wiring back through the original channel and reconnecting the connector. You can now attach the mufflers.

While the car is up on jack stands, start it up and check the oil pressure to ensure that you got the filter on tight. Then let it run for a while. You'll see a lot of white smoke that smells like metal coming from the tailpipes. This is normal and should go away after about 30 minutes of running. You don't need to let it run that long to test for leaks, though. They should be readily apparent.

You want loud? We've got loud—the 2004 C5-R.

SECTION THREE
ELECTRICAL

Projects 18–20

| **PROJECT 18** | Replacing Your Starter |

Time: 1 to 2 hours

Tools: ⅜-inch drive metric sockets

Talent: ★★   **Applicable Years:** 1997–2004   **Tab:** $200 to $400

Tip: Make sure you have a good battery before you replace what seems to be a faulty starter.

Performance Gain: If your Corvette won't start, you're not going to go very fast.

Complementary Work: Think of the battery and starter as a single unit; if the starter fails, make sure you verify the condition of the battery.

Starters seldom go bad. Before you do anything, make sure it's your starter that's really the problem.

Heat Is the Enemy

Every time I hear about a starter problem on a C5 Corvette, I assume that it has long tube headers. Heat is the biggest killer of Corvette starters. General Motors spent a lot of time making sure that the starter didn't fry in normal use. Then you put a set of headers on the car and complain about bad starters.

The problem is called "heat soak." After you turn the engine off, the starter soaks up engine heat and is dead when you turn the key again. Usually, the solenoid is the guilty party, but the wire running between the battery and starter can also be the culprit. To verify that the solenoid is the problem, run the car long enough to get it up to normal operating temperature, then park your Corvette near a garden hose or bucket of water. Turn the ignition off and wait the appropriate amount of time to cause a hot-start problem. When the engine doesn't spin over, try dousing the solenoid with water to cool it and hit the key again. If it cranks over, you have your answer.

You can deal with this heat problem in two ways. The first is to wrap the starter in a thermal blanket that will prevent the heat from reaching the starter. Thermal blankets are available from any speed shop or from a mail order supply company. Most of these blankets simply strap around the starter and are said to reflect around 90 percent of the exhaust heat away from the starter. Thermo Tec takes a slightly different approach and sells a short section of header wrap for the pipes that run closest to the starter. The other technique is to place a metal shield between the starter and the headers to keep the exhaust heat from getting to the starter.

Removal

Removing the starter is pretty normal work, except for one thing—you may have to drop the expensive new exhaust system you just installed. On a stock C5 this isn't necessary. You just knew owning a Corvette was going to be fun.

As with other GM starters, once you remove the electrical connections, simply remove these two bolts and drop the starter down. Keep in mind that you'll go through a lot of batteries before the starter goes bad. But if you use long-tube headers, you can expect some problems from all the heat given off by the header system.

PROJECT 19 | Replacing the Alternator

Time: 1 to 2 hours

Tools: The usual metric combination wrenches and sockets should do just fine

Talent: ★★   **Applicable Years:** 1997–2004   **Tab:** $200 to $400

Tip: If the car has been sitting for a month or more, charge the battery before you try starting the car.

Performance Gain: N/A

Complementary Work: If the alternator fails, replace the battery if it's over three years old.

Corvette alternators get a lot of abuse; far too many Corvette drivers just don't drive their cars enough. You may think that you're preserving your Corvette by not driving it, but you're actually creating a tremendous amount of problems for the battery and the charging system.

A lot of Corvettes get driven only a few times a month. In the North, they might sit idle in the garage for several months. This means that the batteries are very nearly dead when you finally do drive them out of the garage.

You may be able to start the car, but as you drive down the street, the alternator is working harder than it ever imagined possible. The alternator on your Corvette is designed to maintain the battery charge as you drive down the street. It was never designed to revive a totally dead battery.

Fortunately, when GM designed the alternator it built in some excess capacity, but that often isn't enough. The most famous case was the alternator used in the C4 Corvette. Interestingly enough, that very same alternator never caused problems in the rest of the GM product line. Those people actually drove their cars.

When the alternator is asked to work over its design limits, it overheats. Nothing can kill an alternator quicker than heat. This hasn't been a major problem on the C5 Corvette, but you still might want to put a small charger on your battery if the car hasn't been driven in a few weeks. Let the battery charge for a few hours, and then drive your Corvette.

You can also use one of the permanent charging units, such as the Battery Tender (www.batterytender.com). These great little things can be attached to your Corvette while it's in the garage.

A Battery Tender is a 1.25-amp battery charger that automatically switches from full charge to a float-charging mode when the battery accepts only 0.5 amp and then monitors the battery. At that point, the current and voltage of the Battery Tender decrease to 10 milliamps (virtually no current at all) and 13.2 volts. The gassing point of a battery is 13.8 volts; applying any voltage greater than 13.8 for an extended period of time will damage the battery.

The older-style trickle chargers decrease current, but they'll increase voltage to upward of 20 volts and continue the excessive gassing action that's very detrimental to the life of batteries. Battery Tenders are far better for our Corvettes and they only cost around $40.

Three Things You Need to Know About Batteries

There are three important things to look for in a battery: cold cranking amps, reserve capacity, and group size.

Cold cranking amps (CCA) are critical for the battery to crank over your Corvette engine. CCA are a measure of the number of amps a battery can support for 30 seconds at a temperature of zero degrees Fahrenheit until the battery voltage drops to unusable levels. A 12-volt battery with a rating of 600 CCA means the battery will provide 600 amps for 30 seconds at zero degrees before the voltage falls to 7.20 volts. Reserve capacity (RC) is a measure of how long the battery will power your vehicle's electrical system if the alternator fails. The RC

The C5 alternator has been trouble-free.

standard tells you how many minutes the battery can supply 25 amps of power at 80 degrees Fahrenheit without falling below 10.5 volts. The RC standard is determined through a very specific test carried out under standardized conditions.

If you go to your local discount auto parts store, you may see battery ratings called hot cranking amps (HCA) or cranking amps (CA). Ignore these items—these tests are not conducted under standardized conditions. When a battery can't meet the standardized tests for CCA and RC, some companies come up with a test that has no real meaning.

Batteries are subject to a standardized numbering system for their physical size. A battery's group size is a standardized description of its length, width, height, and terminal configuration. All 1997–2004 Corvettes use the same battery size. It's called Group 78, and it uses side terminals.

Battery Load Test

Your local shop can test your Corvette battery's condition with what is called a load test. A traditional load test requires a fully charged battery for accurate results. An otherwise good battery may fail the test if it is not fully charged. The traditional load test applies a specific load to the battery and then monitors the battery's voltage to see if it stays above a certain level. The load created by the testing instrument is adjusted according to the battery's cold cranking amperage.

The load is applied to the battery for 15 seconds while the battery's voltage output is observed. If the battery's voltage remains above 9.6 volts, the battery is good and can be returned to service. If it drops below 9.6 volts, the battery will go dead on you sometime in the very near future. Even if it's close to 9.6, it's not a bad idea to simply replace the battery. I generally replace anything below 10.2 volts just to be on the safe side.

Jump Starting Your Corvette

Jump starting a Corvette is not something you want to do. Every time you do it, you put some very important electrical components at risk. In the old days, jump starting cars was a common practice. But modern electrical components don't like voltage spikes. Make a simple mistake and you could fry the alternator and quite possibly the entire computer system.

When a modern battery is so run-down that it can't crank your Corvette over, there's a good chance that jump starting won't do any good anyway. You might as well replace the battery right there on the spot.

Marine Batteries and Your Corvette

Marine batteries are designed to be almost completely discharged and then brought back up to a full charge. Some folks call them deep-cycle batteries, because going from almost dead to a full charge is called a "deep cycle." There's really no reason to use one of these in your Corvette, unless you seldom drive your C5. Going through a deep cycle has a tremendous effect on the plates inside a normal car battery.

The battery is located directly above your Corvette's computer. If the battery starts to leak and you don't notice it, you'll get to replace both the battery *and* the computer. To avoid problems from a leaking battery, you can install a sealed battery like this one from Optima.

If you have an alternator failure, find a good local shop to test it. An electrical shop can rebuild your alternator to a far-better standard than what you can buy at the local discount parts store. It may take several days, but you can keep your Corvette original and you'll get higher quality at a lower price.

Automotive batteries need to be kept at or near full charge, or the plates can become sulfated and lose their capacity to accept and hold a charge.

A Special C5 Problem

There's one problem that's rather unique to the C5 Corvette. The battery is located directly above the main computer and the associated wiring harnesses. With earlier Corvettes a leaking battery might have been a problem, but with a C5 it's a big problem. The leaking battery acid can cause major expenses. A lot of C5 owners have switched over to an Optima sealed battery to avoid the dreaded leaking battery acid nightmare. These sealed gel cell batteries never leak. When it comes time to replace the battery in your C5 Corvette, consider installing a sealed battery. Also, make sure you check your current battery on a regular basis for signs of leaking acid.

PROJECT 20 | Upgrading Your Audio System

Time: 1 to 2 hours

Tools: You'll not only need the usual wrenches but some special electrical tools such as a wire stripper

Talent: ★★   **Applicable Years:** 1997–2004   **Tab:** $200 to $4,000

Tip: There's no such thing as a simple upgrade here.

Performance Gain: N/A

Complementary Work: N/A

Upgrading a Corvette audio system is becoming more and more difficult. I'm old enough to remember when it was fine to add some extra speakers and a new CD player. But it seems the CD player is going the way of the eight-track tape deck, and you now need a plasma screen, global positioning system (GPS), and a navigation system.

Some Things to Keep in Mind

A lot of us haven't got a clue about what goes on with a modern sound system for the car. That's just one more reason to always keep the basics in mind. Even if you know something about audio systems, it's good to be reminded of the basics.

Quality: Don't expect a set of $30 speakers to sound good. If you truly want to upgrade your Corvette's system, plan on spending significant money for the best components available.

Off-Axis Response: All of the time you're in your Corvette, you're sitting off axis (unless you drive sitting on the console). Sitting off axis affects how you hear sound that is over 3 kHz, a range that humans are particularly sensitive to. The high-end manufacturers take this off-axis seating position into consideration when designing their equipment.

Small-Room Effect: The Corvette is essentially a very small room. Because of this, the nodes and room gain (let's call it car gain) are much higher in this frequency band than at home. Also keep in mind that the car gain is very different depending on vehicles. A Lincoln Navigator has a much different gain than a Honda Civic. This is why you need to find a shop that's done a variety of Corvette sound systems.

Bass Response: There are a lot of factors to clean bass in the car. In the home, room resonance is typically very low, whereas it's much higher in a car. Due to the air volume

You can upgrade your audio system by cutting down on road noise with a sound-dampening system like Dynamat. It's not necessary to cover the entire panel; a few selected pieces will do the job nicely. Remember, the purpose of a sound dampener isn't to absorb sound but rather to cut down on panel vibration.

ELECTRICAL

Most aftermarket head units just don't fit into the dash properly.

restrictions of a car, that air volume plays a heavier role in response and loading. Remember that when in the car, you're actually sitting inside a speaker enclosure.

Imaging: Imaging may be the hardest thing in car audio. Where do you sit in relation to the front stage when listening to music? How about your 5.1 surround-sound home theater? You're generally equidistant from the speakers. A good system will account for the fact you're never an equal distance from the speakers in your Corvette.

Road Noise: No matter what make, or how expensive the model, you always have to consider road noise. The C5 exhibits 93 dB at 65 miles per hour (87 dB at idle) with the top up and the windows up. There are ways to combat road noise, and you'll definitely have to address it. Remember, everything contributes to road noise. We can't do a lot about tire noise and wind noise, but we can do a lot to counter any panel resonance that might occur in the Corvette. Before you spend any money on the actual system, you need to attack the issue of road noise. You can get some great results for less than $500.

Head Unit: The head unit in the C5 is a Delco CD player with a changer control. It has large buttons and plain styling. The Bose speaker system is a little strange (at least compared to other cars), as there is a proprietary connection from the head unit to each speaker. At each speaker there's a small amplifier. This makes upgrades of the rest of the system almost impossible without replacing the entire unit, which you may or may not wish to do. The head unit supports CD and CD-R and has changer control for the factory changer. Note that the changer doesn't fully support CD-R.

One feature of the C5 is the speed-compensated volume (SCV). SCV raises the volume as you drive faster, so that the music appears to have a constant level. This is a wonderful thing if you're driving a convertible, or a C5 with an aftermarket exhaust system.

Amplifiers: Due to the proprietary connection, there's really no way to bench test the small Bose amps. There's a single amp in each door powering the 8-inch midbass. The 3-inch driver is powered from the deck. The Bose amp appears to be a class D design. You can get about 18 watts out of one of the amps, as measured by voltmeter. However, 18 watts probably isn't the maximum, as most measurements are taken during the "best total harmonic distortion at 500 Hz" test.

Speakers: Each door houses an 8-inch midbass and 3-inch full-range driver. The 8-inch driver uses an inverted magnet structure to reduce depth. The lack of a tweeter is apparent in the Corvette system, as the 5-kHz range is very undefined. The 8-inch driver goes into heavy distortion on any signal lower than 40 Hz with any meaningful volume. Overall, the system severely lacks punch. With no tweeter, and running at 15–18 watts per channel, the sound isn't crisp and it lacks detail. The midbass is muddy and vocals are harsh. In other words, it's a pretty cheap system.

The low sound-pressure level, a measure of the intensity of the sound, is limited for the factory audio system. The frequency response was taken at 100 dB. Trying to push past that sound pressure level (SPL) generates a large amount of ringing in the total harmonic distortion tests. When you subtract the road noise level from 100 dB, you get a good idea of your system's dynamic range—not very much. You can generally increase that range by making the volume louder or by making the car quieter. This puts the C5 owner who installed a loud muffler in a real quandary.

What Features Do You Really Want?

CD: Almost every head unit will play CDs, so listen to some of your favorite CDs to get a feel of the noise floor of CD playback. (Noise floor means how much noise you get when there shouldn't be any noise at all.) Some players have techniques to improve the perceptible noise in CD material.

DVD: Even if you aren't considering watching movies in your Corvette, keep in mind that DVDs can provide a much better signal and a lower noise floor than CDs.

Multimedia: There are some really cool things out now for cars, especially if you have kids. Dual-zone systems allow you to listen to a CD while your passenger watches a DVD. Better yet, hook up an X-box, and let your passenger play video games while you drive. Multimedia systems allow other video inputs, such as satellite navigation or a camera for seeing behind you when you are in reverse.

MP3: While not the best from a quality standpoint, MP3s are absolutely the best from a quantity standpoint. You can fit 200 songs on a CD in MP3 format, eliminating the need to carry around all those bulky CD cases. If you want great sound with your MP3s, you can rip your own music from your CDs at 192 kb/s or higher.

Sound Control: With the Corvette you need as much control over the sound as you can afford. This is usually done with an equalizer (EQ) function. Seven to 10 bands

A huge bass speaker like this one gets instant street respect. The bass speakers in the C5 Corvette are actually pretty good; it's the lack of a decent tweeter that's the biggest problem with the C5 audio system.

In the stock Bose system, each speaker has its own amplifier, so there's no such thing as a simple upgrade. You either live with what you have or you purchase a whole new system. Your money is better spent on silencing road noise.

are common, but to really fine-tune the system, you need a 15- or 30-band equalizer. Some higher-end decks now come standard with time alignment, which enables you to compensate for off-axis seating. Time alignment is one of the most important features to look for in a modern car sound system.

Expandability: A lot of the newer decks feature expandable network buses that allow you to add processors, satellite radio, CD/DVD changers, and zone switchers, all in a single shielded wire. Each manufacturer has its own bus name and proprietary connection/transport.

Processor: If you want to get the best sound and best tunability, you'll need a processor. Look for 30 bands of EQ per channel, 5.1 decoding (if you want that), time alignment, and digital inputs. Even better is having the ability to store a few profiles for different music and seating positions. In-car audio digital inputs tend to be proprietary, so double-check the connectors when you are buying the various cables.

Amplifiers: In the entire car audio industry, this is the area that turns technology into religion. There are some very reputable amplifier manufacturers out there, but also some major scam artists. Stick to major brands or brands that have very good reputations. Look for reviews in reputable magazines and listen to the amps yourself. It's very common to see an amplifier that is 8x8 inches in size claim to output 1,000 watts at less than 1 percent THD—yeah, maybe if it got struck by lightning.

Speakers: As in home audio, listen, listen, and listen to speakers before you buy them! Speakers are personal taste, but here are a few pointers:

Positioning: When the speakers are installed, they will most likely be off axis, so listen to them off axis (in the driver's seat). Most audio stores position the speakers on a shelf and stand you directly in the middle of them, but that's certainly not the way you'll listen to them in the car.

Also keep in mind that in the Corvette your ears are above the speaker system. Most stores position the speakers in perfect locations. Ask them to move the speakers around to resemble the way you'd be listening to them in your Corvette.

Power: Stores have a habit of playing speakers for you with big-power amplifiers, and this leads to great detail and a solid sound. Unfortunately, if you aren't going to power the speakers in your Corvette with similar power levels, the speakers will sound very different. Ask to listen to them at your level of power. Make sure you find out what kind of power they're using in the store, and then ask if they could be hooked up to something that resembles the level in your Corvette.

Price: The same manufacturer will usually have three or four price levels of speakers. Much of the technology is common between the lines, so listen to all the lines if you like a particular manufacturer. I find that the difference between the top-of-the-line speakers and the midpriced speakers is often minimal.

Road Noise, Again

With the C5 Corvette there isn't much you can do to upgrade the audio system due to all the special connections. This means installing an entirely new system in your Corvette, which most of us don't want to do. If you can lower all the other noises, such as road noise, the C5 system will sound better.

The biggest difference between your Corvette and your home stereo is the background noise. There are all kinds of exterior (road noise, rain hitting the windshield, etc.) and interior (rattles) noises that draw attention away from the music in your Corvette. To make up for the road noise, we usually turn the stereo up louder.

While it's impossible to eliminate the noise completely in your Corvette, there are products that will decrease the

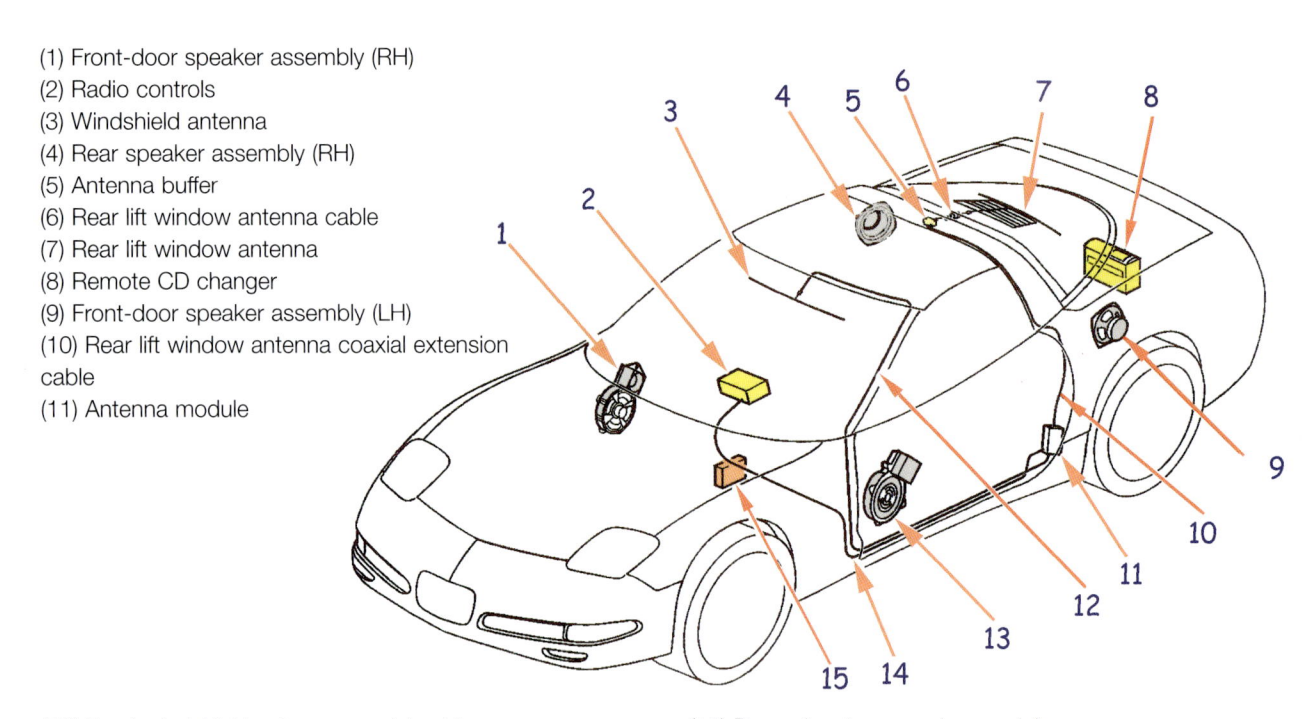

(1) Front-door speaker assembly (RH)
(2) Radio controls
(3) Windshield antenna
(4) Rear speaker assembly (RH)
(5) Antenna buffer
(6) Rear lift window antenna cable
(7) Rear lift window antenna
(8) Remote CD changer
(9) Front-door speaker assembly (LH)
(10) Rear lift window antenna coaxial extension cable
(11) Antenna module

(12) Front windshield antenna coaxial cable
(13) Front-door speaker assembly (LH)
(14) Radio control coaxial cable

(15) Bose signal-processing module
Courtesy of General Motors

noise a great deal. Reducing the noise in your Corvette will make a big difference in the audio system's performance and overall ride comfort.

Sound-Dampening Liners

You can cut down road noise dramatically by installing a sound-dampening product like Dynamat. These dampeners work by absorbing sound-causing vibration energy, eliminating speaker resonance, and baffling out excessive sound. And since they work as converters rather than simple noise blockades, a little goes a long way, saving you money and installation time.

To add a mat liner to your Corvette, the seats, carpet, door panels, etc., have to be removed. Using a heat gun and a small wallpaper roller, the material can be laid over door panels, floors, wheel wells, etc. A cheaper alternative to Dynamat is a product used by roofing contractors called "Ice Guard," which has an adhesive backing and works the same way.

Sprays

There are noise-dampening products that can be sprayed onto the panels, such as Rockford Fosgate's Noise Killer Blue. Sprays are used in places where a liner can't be applied, such as inside doors, the rear compartment, etc. Most of these products are either sprayed on or applied with a brush. Some sprays need an air compressor and a spray nozzle, and others, such as Stinger RoadKill (www.stingerelectronics.com), already come in a spray bottle.

You may instead apply a rubberized undercoating that can be obtained at any major car parts store. It comes in a spray can and is easy to apply. The only drawback is that the undercoating is very sticky and messy.

Expandable insulation spray foam is used in homes to seal around pipes and fill up holes in basements. In a car, it can be used on irregular surfaces where tar mats can't be applied, such as the trunk, trunk lid, etc. Once the foam dries (about 4 hours), simply cut the excess off with a long knife.

Adhesive Strips

Adhesive strips are normally used for home door insulation. The strip of foam has an adhesive material on one side and is used to seal between the door and the doorjamb to keep air from escaping the house. Apply these strips between panels, behind license plates, etc., for a quick, inexpensive, and easy way to get rid of annoying rattles.

Sound deadener may well be the best investment you can make to improve the sound quality of your Corvette. For every 3 dB of road noise you block, you effectively make your sound system 3 dB louder. Actually, the key is that you can turn the system down 3 dB and still hear what you were hearing before. This means an easier job for the amplifier and better dynamics. Decreasing road noise also provides a chance to eliminate rattles and loose items. If you use a variety of products, you can probably reduce the road noise by almost 6 dB.

SECTION FOUR
DRIVETRAIN

Projects 21–26

PROJECT 21	Installing a Short-Throw Shifter

Time: 1 hour

Tools: There's no need for any specialized tools

Talent: ★★　　　**Applicable Years:** 1997–2004　　　**Tab:** $200

Tip: N/A

Performance Gain: Theoretically, you might save a few hundredths of a second shifting from one gear to the next.

Complementary Work: Make sure that there is synthetic fluid in the transmission if shifting is important enough to install a new shifter. Even though the car came with synthetic fluid, it could benefit from new fluid.

Installing a short-throw shifter is easy, and just as easy to reverse if you decide it's not for you. Or, more importantly, when the lease on your Corvette expires or you want to sell the car, these shifters can be easily removed. The average person is a little suspect of a used Corvette that has been modified for extra performance.

I've installed short shifters in my car and ended up wondering if it was worth the trouble, but a huge number of people swear by this change. If you're serious about drag racing, it might be a good idea, but I've never seen anyone get quicker elapsed times because of a new shifter handle.

It comes down to the fact that a short-throw shifter feels cool. If you purchase a kit with a billet handle it also looks pretty cool. That may be reason enough to have one in your C5.

There are a variety of shifters on the market, but the directions are basically the same. You start by using a small

You can adjust the lever of this aftermarket shifter to a variety of positions.

73

This shifter is in a C5 race car.

flat-blade screwdriver to remove the plastic shift pattern from the knob. This exposes a square pin that holds the shift knob on securely. The hardest part of this whole project might be removing the square pin in the shifter, depending on how hard the assembly line worker hammered the pin into place.

With the little wedge out of the way you can open the console, and use a screwdriver to gently pry up the trim piece with the traction control button. Disconnect the traction control harness and the variable suspension selector harness (if your car has one). You may also need to disable the air bag disable switch if your car has one. Finally, disconnect the accessory plug wiring harness and remove the two 10-millimeter nuts located under the trim piece you just removed.

Gently pry off the plastic covers and remove the two 10-millimeter nuts located inside and at the rear of the console. Now you're ready to pry off the interior temperature sensor and remove the T15 Torx screw.

At this point you can remove the ashtray. Remove the T15 Torx screw that's behind the ashtray and the T15 Torx screw that you'll find alongside the cigarette lighter.

If you have a convertible, you'll have to remove the waterfall panel between the seats. Now you should be able to slide the console toward the rear of the car. After removing the console/dash trim plate and disconnecting the cigarette lighter connector, you can remove the waterfall. Do this by removing a T15 Torx screw on the top and two T15 Torx screws at the base of the waterfall. Lift straight up and set it aside in a safe place.

With the shifter exposed, you can remove the four 10-millimeter nuts holding the shifter boot down to expose the shifter base. Then remove the four 10-millimeter nuts from the shifter base and lift the old shifter out. If your kit came with a new gasket, scrape off the old gasket with a razor blade. Don't forget to pop off the black plastic insulator.

You can finally install the new shifter and put everything back together. This is basically a reversal of the disassembly with the exception of installing new parts and/or gaskets supplied with the shifter.

Don't forget to replace the black plastic insulator on the ball of the new shifter. Install the new shifter using a new gasket, and tighten the four nuts that hold the shifter base. Tighten these carefully but don't strip them. Now reinstall the boot and the four nuts. This is a good time to check that you can shift into all the gears.

If everything moves properly, you can install the console trim. You may have to slide the console compartment back again to get everything to fit. Don't forget to connect the cigarette lighter connector and then check to make sure it works.

Slide the console compartment forward and install the four Torx screws you removed earlier. Remember that two go under the traction control switch and two inside the console. Now turn to the nut covers inside the console and replace the inside temperature sensor cover. If you're working on a convertible, you'll have to reinstall the waterfall.

It's now time to connect all the electrical connections that you disconnected earlier and replace the traction control panel. If you're installing a new shift knob, follow the manufacturer's instructions. If you're replacing the stock knob, screw it back on and then insert the square pin back in its hole. This pin should go back in place while gently tapping it into the knob. Don't pound the pin into the hole. The last step is to replace the plastic shift pattern.

Test-drive the car to make sure everything works. Make sure you also test the traction control variable suspension and the air bag disable switch.

To remove the old shifter, first remove the plastic cover and then you can remove the plug that was hammered into place. In some Corvettes it's easy, and in others it'll make you crazy. *Courtesy of General Motors*

PROJECT 22 | Servicing the Six-Speed Manual Transmission

Time: 1 hour

Tools: This can actually be done with a large Crescent wrench if you lack the open-end metric tools.

Talent: ★★ **Applicable Years:** 1997–2004 **Tab:** $25 to $50

Tip: Be sure that you check for leaks while you're under the car.

Performance Gain: You might notice improved shifting.

Complementary Work: N/A

It doesn't get any easier than this, but you will have to unplug the temperature sensor to check the fluid level. GM uses the temperature sensor as a fill plug. The drain is on the lower right-hand side. It's a conventional drain plug that you can remove with a wrench. As easy as this job is, there's no reason not to do it annually.

I recommend using a synthetic transmission fluid in this unit simply because of heat issues. GM went to the trouble and expense of installing a temperature sensor in the transmission, so we know that heat is an issue.

Because of the front air dam and the smooth undercar aerodynamics, it seems there's limited transmission and rear axle cooling. All of the things that improve the gas mileage on the car also keep air from flowing around the transmission and rear differential. This causes things to get very hot. The manual transmission temperature sensor turns on the "trans over temp" light when the transmission oil gets above 325 degrees Fahrenheit. That's why GM has stated, "the transmission could be damaged if not allowed to cool down."

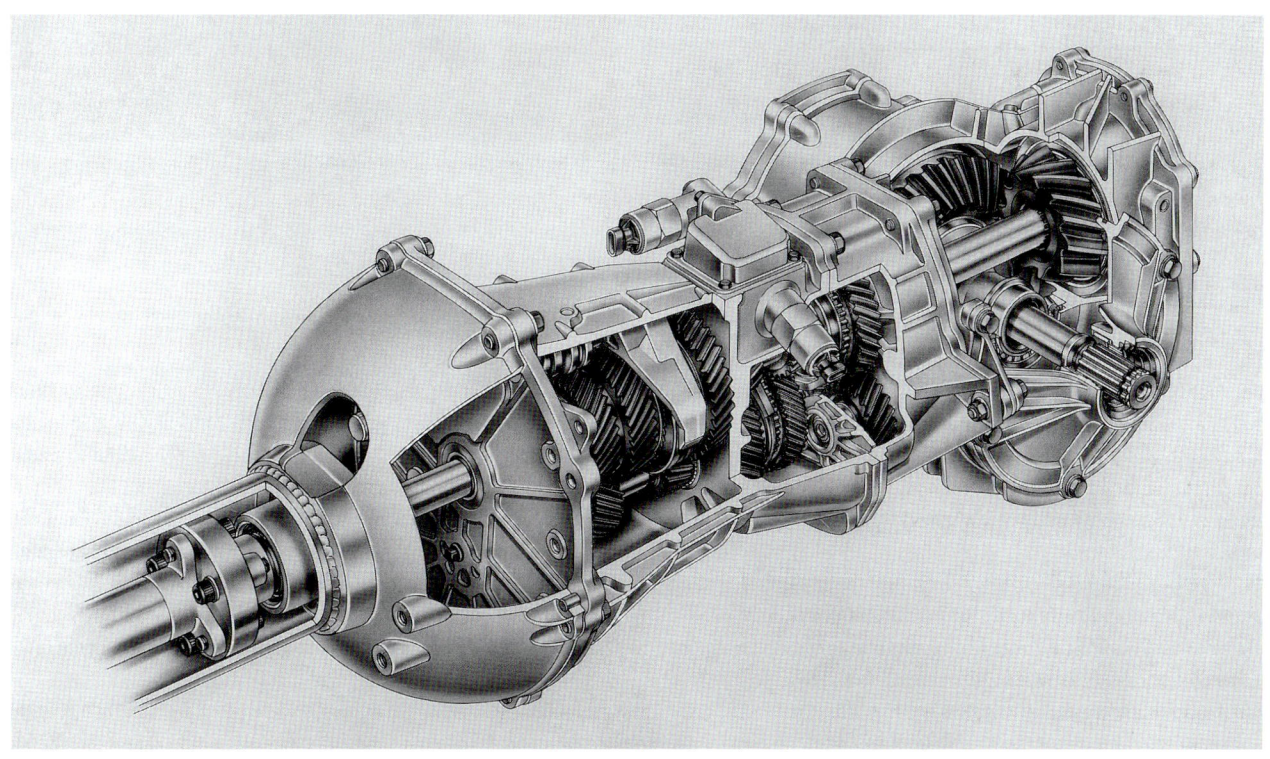

Courtesy of General Motors

Transmissions	Hydra-matic 4L60-E	Tremec T56 (MM6)	Tremec T56 (M12)
Type	four-speed automatic	six-speed manual	six-speed manual
Application	standard on coupe and convertible	optional on coupe and convertible	standard on Z06
Gear Ratios			
First	3.06	2.66	2.97
Second	1.63	1.78	2.07
Third	1.00	1.30	1.43
Fourth	0.70	1.00	1.00
Fifth	n/a	0.74	0.84
Sixth	n/a	0.50	0.56
Reverse	2.29	2.90	3.28
Final drive ratio	standard: 2.73:1 optional: 3.15:1	3.42:1	3.42:1

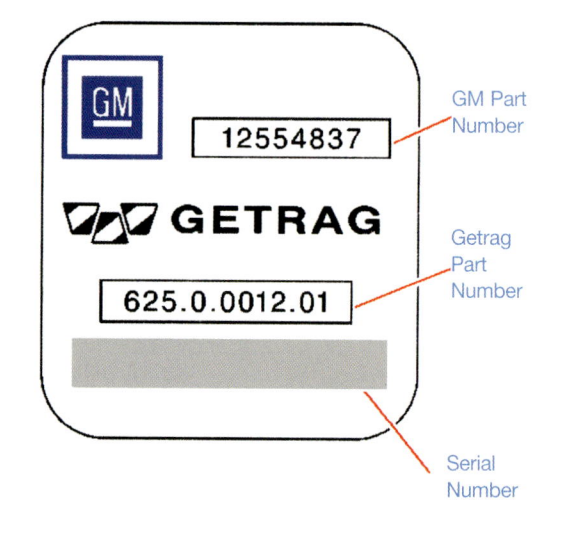

This tag, attached to your six-speed transmission, carries all of the relevant data and can be used to determine if you still have the original transmission. *Courtesy of General Motors*

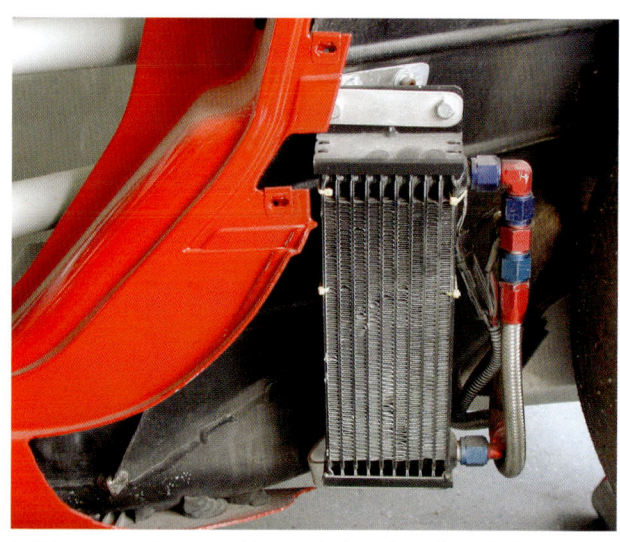

In this race car's manual transmission, the oil cooler has been placed behind the Z06 rear brake duct. This is not the stock location. Once the body panels are in place, it's hidden from view and it's also impervious to rocks and debris.

It would be better to have a transmission oil cooler. GM makes a transmission cooler and it's available from your local Chevrolet parts department. Most Corvette owners will never set off the light, which is why GM left the cooler off the standard car. But, if you run track events, consider having a rear axle cooler installed.

The C5 Corvette uses the Tremec T56 six-speed manual transmission, derived from a transmission previously used in the Camaro. GM says that no maintenance fluid change for the C5 manual will ever be needed, but there is nothing wrong with new gear oil once every few years.

Check the fluid level with the engine off. Remove the transmission fill plug. If the level is even with the plug hole threads or fluid dribbles out, your transmission is full. If fluid leaks profusely, it was overfull and you should allow the excess to leak into a drain pan. If the level is below the hole threads, add fluid until it's even with the threads.

To drain the fluid, remove the T56's drain plug. When the fluid stops running, replace the drain plug and refill the gearbox through the fill plug hole. The fluid in the T56 is Dexron III ATF. Mobil 1 would seem to be the obvious choice here, but I doubt if the brand really makes that much difference.

The Clutch

The clutch is operated with a basic hydraulic system. When you push down on the clutch pedal, pressure is applied to

GM tried to make it pretty obvious about the correct transmission lubricant. Just think—in about 30 years someone is going to be reproducing these labels.

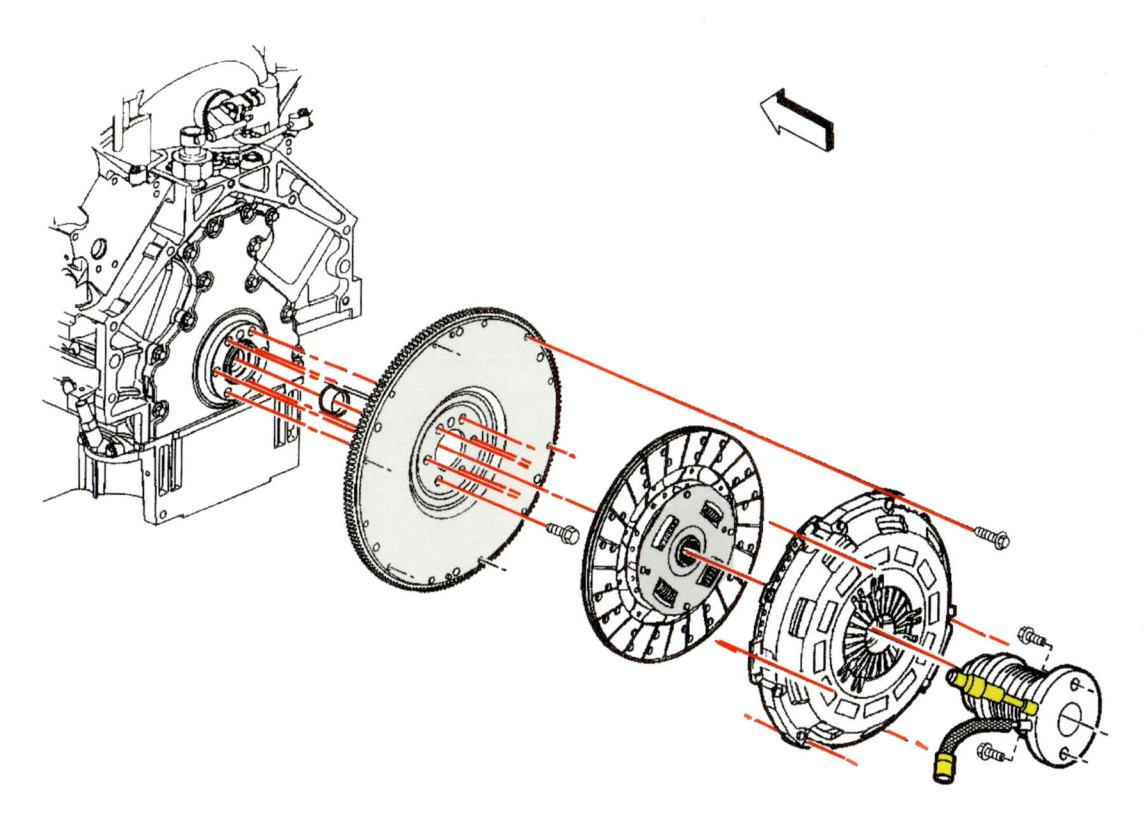

Even though the transmission is in the rear of the Corvette, the clutch and flywheel are in the traditional position directly behind the engine. *Courtesy of General Motors*

the plunger in the master cylinder. As the fluid moves it actuates the piston in the slave cylinder, forcing the release bearing to disengage the clutch pressure plate from the clutch disc.

Aluminum Flywheels

If you're into performance, consider installing aluminum flywheels. If you use your Corvette on the track for driving events, installing aluminum flywheels may be the best single modification you can make.

This isn't a huge job, but it does involve removing the transmission—something you may not want to tackle at home.

Aluminum flywheels won't give you any more horsepower but they will let you get up to serious horsepower a lot quicker. The idea is to exit the corner quickly and get to the next corner before the other people.

77

PROJECT 23 | Servicing the Automatic Transmission (Fluid Change)

Time: 1 hour

Tools: ½-inch drive metric sockets

Talent: ★★★★   **Applicable Years:** 1984–1996   **Tab:** $50

Tip: Make sure you ask a lot of questions before you let a shop tear into your transmission. The expense is great, and a lot of things can happen to your car while it's in the shop. Make sure you find a top-quality shop. The quality of the work will be a lot more important than the final price.

Performance Gain: N/A

Complementary Work: If you're considering installing a shift kit in your automatic, keep in mind that you'll be changing the fluid and filter at that time.

Most Corvette owners prefer the automatic transmission to the manual transmission. Very little ever goes wrong with the automatics; they are very solid transmissions and seldom need repair. The most common reason for repair is simple abuse; the most damaging form of abuse is neglect.

Since many of you purchased your C5 Corvette on the used-car market, you need to check out the transmission. You should check the fluid level with the engine running when the car is completely warmed up.

Checking the Fluid

The C5 Corvette uses GM Hydra-matic Division's 4L60-E automatic transmission as base equipment. It was strengthened over the C4's 4L60-E transmission with a more robust torque

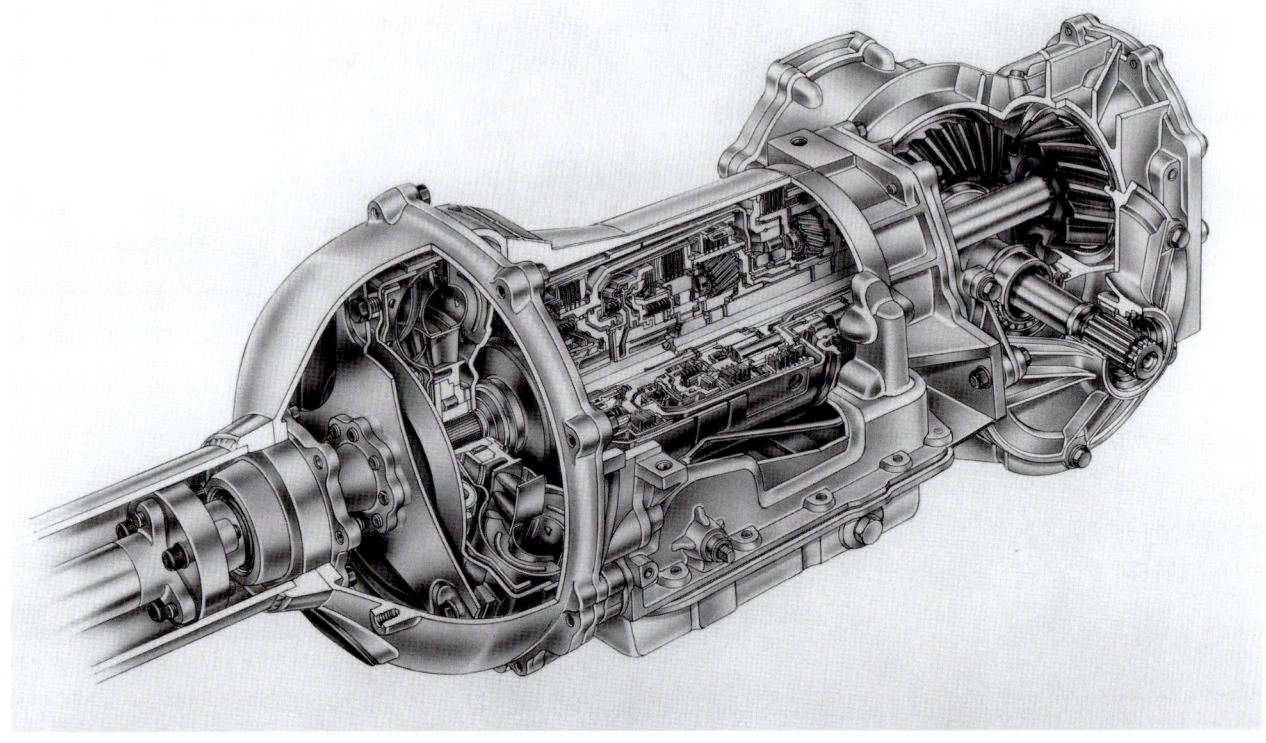

Courtesy of General Motors

The transmission drain pan on the C5 is very conventional. The mess in your garage will also be very conventional. Be prepared to have transmission fluid all over you and your garage floor.

converter, a new fluid pan for more consistent fluid pressure during high lateral acceleration, and revised calibration. Wide-open-throttle (WOT) shifts occur at 6,000 rpm. To improve durability, transmission controls include torque management, which retards timing for a fraction of a second during WOT shifts to reduce shock loads.

Unlike previous Corvettes, the C5 has no dip stick. Fluid checks are accomplished while working under the car with the engine running. This may involve hot fluid, powertrain, and exhaust parts, so we suggest wearing a pair of gloves while you do this.

If you're going to check the transmission with the car on jack stands, make sure it's high enough for you to safely move around while the exhaust system is hot. Start the engine and, using the "gauges" button on the driver information center (DIC), put transmission temperature on the dash display. Run the engine until the trans temp is between 86 and 122 degrees Fahrenheit.

Apply the brakes and shift into each gear for a few seconds, and then shift back to park. Set the parking brake, leave the engine running, and crawl under the car. Position a drain pan under the fill plug area and remove the plug. If fluid dribbles out or is right at the bottom of the fill hole, your transmission is full. If fluid leaks profusely, it was overfull and you'll need to allow the excess automatic transmission fluid (ATF) to drain into the pan. If no fluid comes out, you will need to add more in a minute.

A small screwdriver, a hex key, or even your little finger works nicely as a dip stick to inspect the fluid color. I like to put some drops of the old fluid on a white paper towel. Then I put brand-new fluid drops next to it to get an idea of how the old fluid is doing.

Red, or red with a light-brown tinge, is a normal color. If the fluid has a medium-to-dark-brown color or smells burned, change the fluid and filter. If the fluid has a cloudy or milky appearance, it's contaminated with coolant. The solution to coolant contamination is beyond the scope of an article on basic maintenance. See the service manual for further information. If you can see clutch material on the paper towel, you can be almost certain that the clutches need to be replaced.

If you need more fluid, add enough to bring the level up even with the bottom of the fill hole; replace the transmission fluid fill plug and shut off the engine. If you added a large amount of fluid, there is a leak somewhere that needs to be repaired. Start by cleaning the transmission case with brake cleaner, and then check a day later to see where it's leaking.

GM says that the Corvette has a 100,000-mile ATF and filter change interval; however, there is an exception to that noted in the service manual. If your C5 is regularly operated in heavy traffic when the ambient temperature is above 90 degrees Fahrenheit, or operated in hilly or mountainous

Disconnect the transmission fluid temperature sensor electrical connector so that you can remove the fill plug. *Courtesy of General Motors*

Once the new filter is in place, position the oil pan and tighten the bolts alternately and evenly to 11 N-m (97 in-lb). *Courtesy of General Motors*

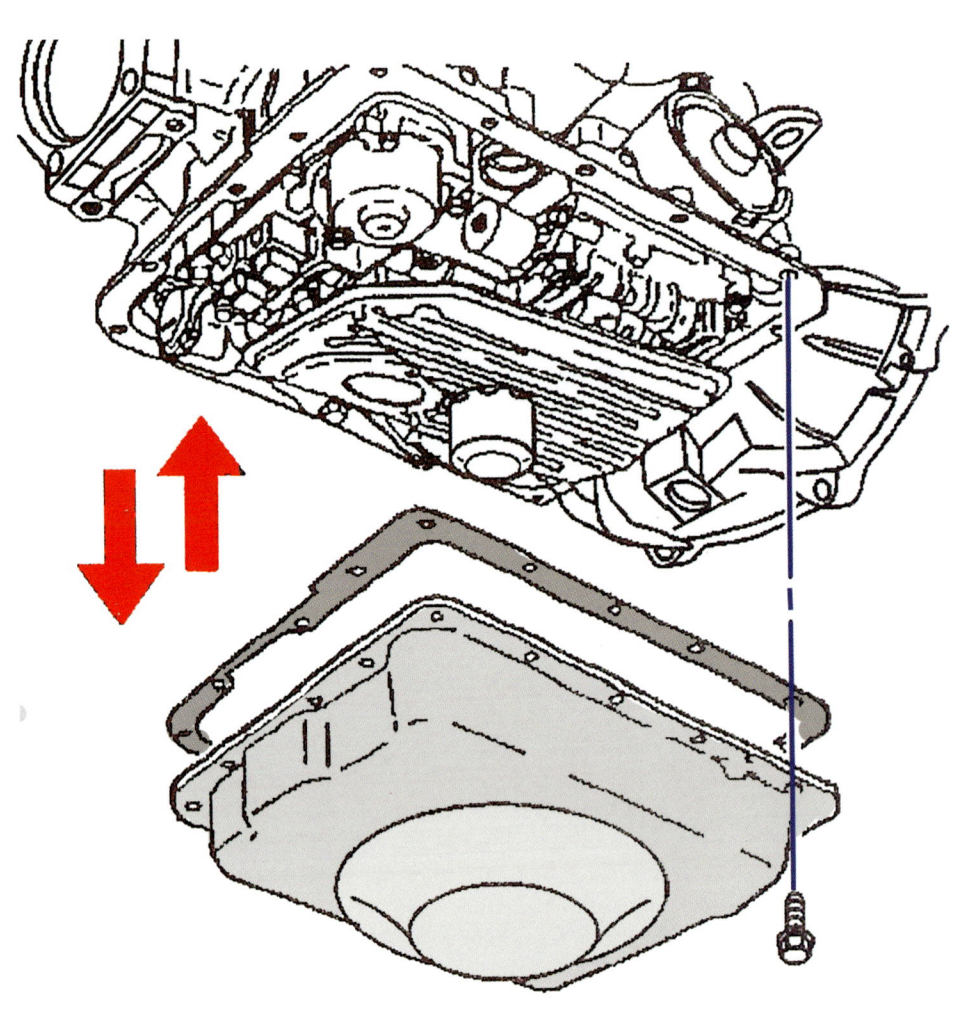

areas, or operated in a high-performance duty cycle, the fluid/filter change interval drops to 50,000 miles. The enemy of any automatic trans is high fluid temperatures. If peak fluid temperatures are reduced, fluid life increases. To be on the safe side, I always go with the 50,000-mile interval.

A fluid and filter change in a C5 is no more difficult than in Corvettes of the past, although it's just as messy. Support the car safely, place a drain pan under the transmission, and remove the fill plug. I recommend that you put some large plastic sheeting on the garage floor the width of the entire car and up to 3 feet in front of and behind the transmission. Just assume you're going to make a mess. It wouldn't hurt to have a bag of kitty litter handy as well.

Don't forget proper lighting. It's dark under your Corvette, and you want to make sure that you see everything that's going on. This is one case where those inexpensive halogen lights come in very handy to focus the light directly into the correct area.

With the engine off, the transmission pump isn't circulating fluid, so there will be significant drainage out the fill hole. That means it's going to drip right down the side. Once the dripping stops, support the transmission pan and remove its bolts. Carefully remove the pan and drain the remaining ATF. A rawhide hammer works nicely to break the pan loose.

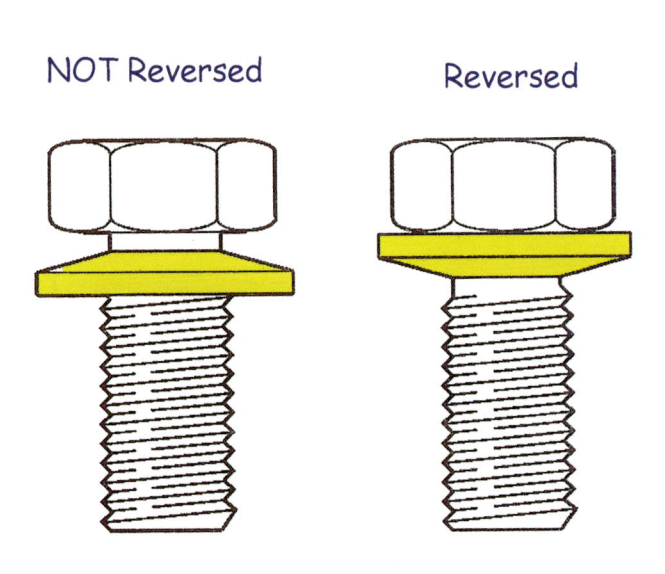

Inspect all of the oil pan bolts and washers to determine if the washers are reversed. Reuse the bolts and washers if the conical washers are not reversed. Replace any of the oil pan bolts and washers that are reversed. *Courtesy of General Motors*

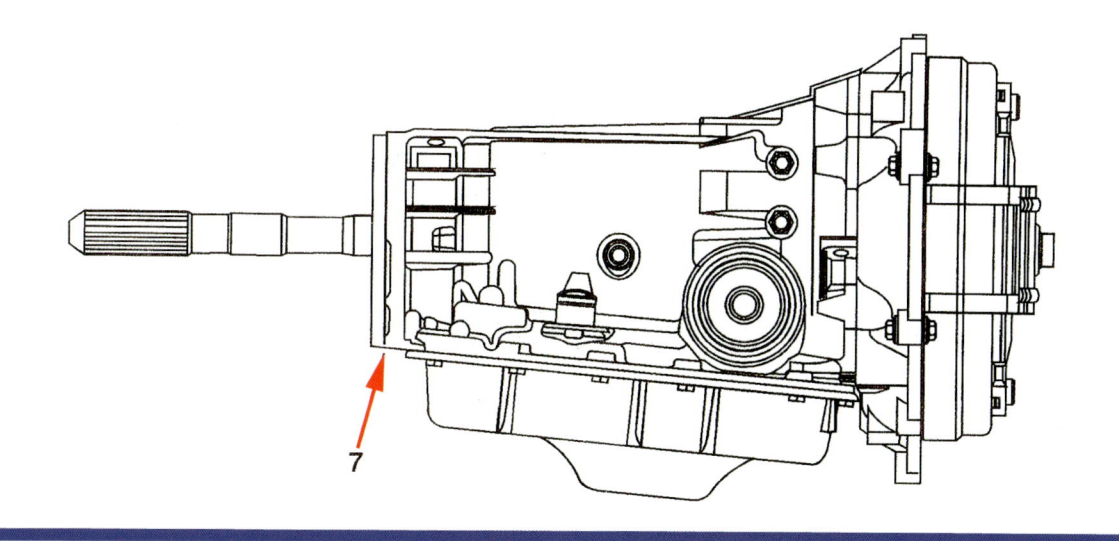

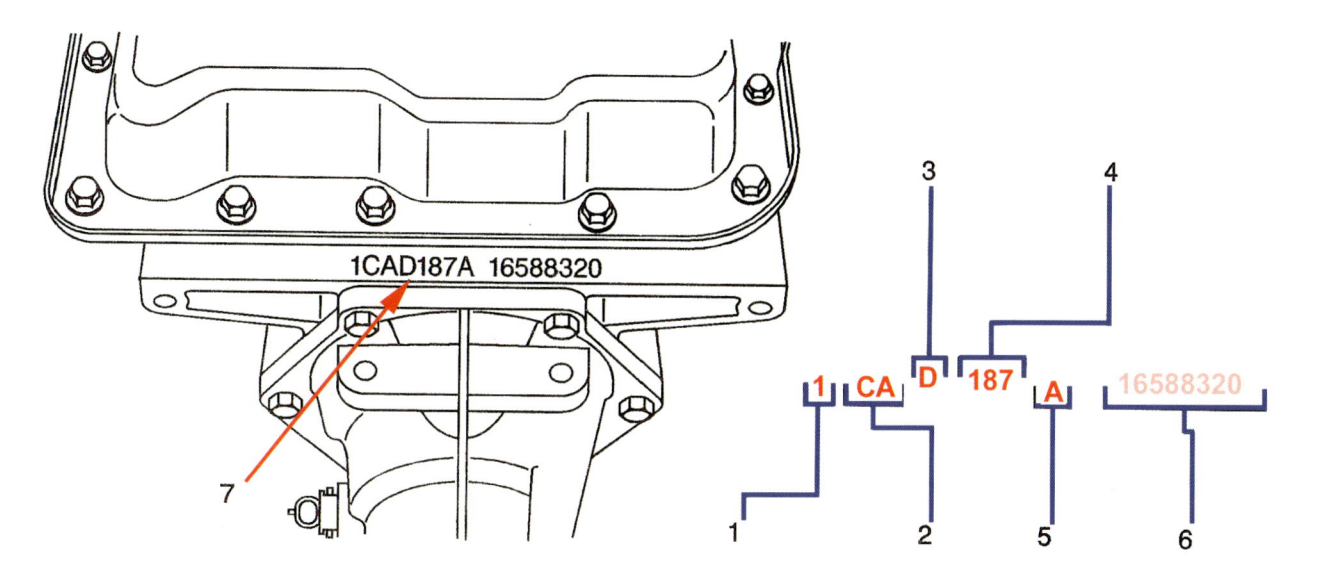

(1) 1 = 2001
(2) Model
(3) Hydra-matic 4L60-E
(4) Julian date
(5) Shift built (A, B, J = first shift; C, H, W = second shift)
(6) Serial number
(7) Transmission ID location
Courtesy of General Motors

Once the pan is off, carefully inspect its bottom. A silver or gray residue is normal; however, a significant amount means the transmission has a problem. A "significant amount" is hard to characterize, but the old rule has always been, "If you see more than a dime's worth of metal in the bottom of the pan, overhaul it." Whether it's a dime's worth or a significant amount, if you see a lot of metal in the transmission pan, put it back on, refill the trans, and take the car to a transmission service facility for diagnosis and possible repair.

If there's just a normal amount of sediment, set the pan aside for cleaning and take a look at the filter. Note the position of the old filter before you start installing the new filter. You can remove the filter by gripping it firmly and pulling down while twisting a bit. Throw both the filter and filter seal away. If the seal is stuck up inside the valve body, use pliers to pull it out. Make sure you remove all traces of the old gasket from the transmission case and pan.

I prefer using an ACDelco transmission filter kit (TF306). A variety of pan gasket and filter sets are available at your local discount parts house. They may be either boxed or in flat shrink-wrap packages. Flat shrink-wraps are typically used with cork pan gaskets to protect the gasket from bending. Rubber (neoprene) pan gaskets, on the other hand, are flexible and can be folded to fit in a box. Different technicians have different preferences as to which is the best gasket material to use.

Silicone gasket material that comes in a tube can cause you more problems than you can imagine. When you use

The shift lever in the C5 uses a cable system of actuation.

silicone gasket material, it's squeezed between the transmission body and the pan. You'll notice how it squeezes out of the sealing faces; just as much has been squeezed into the oil pan area. If this material gets loose inside the pan, there's a good chance it will find its way into internals of the transmission and clog a very important passage.

In addition to the pan and gasket set, you'll need 4 or more quarts of Dexron III ATF to refill the transmission. The transmission's fluid capacity is important—automatics must be run very close to the full level but never over that level. Overfilling your automatic transmission can aerate the fluid, causing shifting problems. Underfilling can cause slow engagement and slipping. The amount of ATF required is usually listed in your owner's manual.

Three different torque converters were used during the production of the C5, meaning the fluid capacity can be anywhere from almost 9 quarts to just over 12 quarts. Keep in mind that this measurement assumes that the torque converter has been drained dry, which is generally not the case when you're simply changing the filter. The best solution is to ask the dealer how much Dexron III you need to use after changing the filter.

Install the new filter seal in the valve body and push the new filter in place. Clean out the transmission pan and look closely at its bolt washers. The flat sides go toward the pan. If they are reversed, get new bolts. If they're not reversed, reuse the bolts. Apply the new gasket to the pan. Install the pan and torque the bolts to 8 ft-lb.

Add fluid until it leaks out and then install the fill plug. Perform the fluid check procedure the same way you did earlier and, if necessary, add more fluid. Since 1994, the factory-fill fluid in Corvette automatics has been an organic-based ATF that meets the Dexron III specification. Mobil 1 synthetic fluid works well too.

Some shops use transmission service equipment that can replace the fluid without dropping the transmission pan. This equipment either taps into the ATF oil cooler lines or connects to the filler tube. But this service doesn't include a filter change. The only way to change the filter is to drop the oil pan. I'm old fashioned enough to think you should be more concerned about changing the filter than you are about changing the fluid.

Additives
A lot of additives are sold on the market for transmissions, and none of them are good for your transmission. The only time I would use an additive is in a moment of desperation. Case in point: When the seals have started to leak and you want to postpone any real repair, you may use an additive with chemicals that swell the seals and gaskets to reduce fluid leakage. This might hold you over until you can afford to have your transmission properly repaired, but don't think of the additive as a real repair.

Getting Your Transmission Back
Carefully explain to the shop that you want your transmission rebuilt, not simply a rebuilt transmission. The large transmission shops keep a variety of units on the shelf to offer same-day service on transmission repairs. To be on the safe side, crawl under the car with a light and locate the VIN number on your transmission. This VIN number can be in more than half a dozen locations, so check carefully. Then when you get your car back home, check this number again. This might not seem important to you right now, but at some point having the original driveline in your Corvette will be a big deal. Just ask the people that own 1968 and 1972 Corvettes how much they would be willing to pay to get the original transmission back in their Corvettes. Don't neglect it now.

PROJECT 24 | Selecting a Shift Kit for the Automatic

Time: 1 day

Tools: All of the usual transmission tools, plus an assortment of ¼-inch drive sockets and very good lighting

Talent: ★★★★★ **Applicable Years:** 1997–2004 **Tab:** $150

Tip: Make sure you don't get a shift kit that is too aggressive for a street-driven Corvette.

Performance Gain: The shifts will be a little quicker at wide-open throttle, but at the same time will feel slightly harsh driving around town.

Complementary Work: Change the transmission fluid and filter at the same time the shift kit is installed.

Shift kits work by altering the control pressure and oil flow within the transmission. They usually include a redesigned control plate and different gaskets that mount between the valve body and transmission housing. The new plate may have different orifice sizes; relocated, new, or blocked ports; and may come with recalibrated control springs. Shift kits are not necessary for most street-driven Corvettes. GM has already given the Corvette crisper shifts than the other cars in its product line.

A shift kit generally makes changes to the valve body in order to firm up the shifts. Not all shift kits improve shift firmness a great deal; some are made specifically to correct (and prevent) problems that can wreak havoc on the transmission later on.

Even though the Corvette transmission is the same as that found in the Camaro, the transmissions work differently. A number of internal parts are Corvette-specific and are not used for the rest of the GM product line. Many Camaro owners love to put the Corvette servos into their Camaros.

Some of the high-performance aftermarket shift kits can actually change the operation of the transmission. They alter the valve body significantly enough to change how the transmission behaves. For example, several high-performance kits can convert an automatic transmission to have fully manual shifts, converting the transmission into a clutchless standard transmission.

The crux of all this is that improving your Corvette's transmission is not as easy as simply purchasing a shift kit. Shift kits don't come in a standard size, and very few people have the knowledge necessary to advise you on your choices.

Skilled transmission technicians are extremely rare. The good ones all have six-figure incomes. These are also the only people you should listen to if you're considering installing a shift kit. They can get you pointed in the right direction and install the kit for you.

When dealing with your automatic transmission you can also have the electronic controls reprogrammed by a shop with a programmer like Tech 1, or this can even be done with the Hypertech Power Controller if you want to do it yourself. With the Power Programmer III, you can electronically increase automatic transmission line pressure for firmer shifts while reducing clutch slippage and transmission oil temperatures. The increase in line pressure will give you crisper (harsher) shifts than you get from the standard programming.

You can also adjust automatic transmission shift points for maximum performance for quicker acceleration and better quarter-mile times, as well as adjust the part-throttle shifting to work properly with nonstock tire sizes and/or nonstock rear end gear ratios.

The Corvette's computer calculates your speed based on the stock tire size and rear end gear ratio. If you install taller or shorter tires, or change the rear end gear ratio, the vehicle computer doesn't know how fast your car is really going. It may signal the transmission to upshift too soon or to hold a gear too long. With the Power Programmer III or the Tech 1, you have the ability to customize the shift points in every gear. This programming allows you to change back to the stock setting if you decide you really don't like the harsher shifts.

Factory transmissions intentionally soften up the shifting for various reasons by allowing a generous amount of "slippage" in the transmission. This is what causes wear on the clutch and bands. Remember, though, that a hand-held programmer cannot mechanically alter the way the transmission operates. It can boost the maximum line pressure across the board, but that can harm the transmission if run that way for an extended period of time. The computer allows you to boost the line pressure at the track and then bring it back down if you're looking to avoid damage on the way home.

DRIVETRAIN

The automatic transmission is very conventional, even if it is in the back of the Corvette. It was placed here to free up more interior room. It has no dip stick for checking the level of the fluid. Instead, you have to get the car on a lift or jack stands and check the fluid level by removing the fill plug.

The Next Step

After you accumulate some miles with the new electronic program, you need to think about whether you want to go even further. The next step is to have your transmission shop install something like the TransGo shift kit from Transmission Exchange Co. This kit is installed inside the transmission and mechanically alters the way the transmission oil flows within the valve body by replacing five valves and 11 springs, and actually only boosts line pressure minimally.

The TransGo kit provides improved shifting. The 1–2 shift firmness can be isolated and mechanically altered by adjusting the pressure in the second accumulator by shimming the spring seat. The band is mechanically adjusted for the correct amount of play by shimming the second piston housing; the reverse manual valve is chamfered to provide smoother, faster release.

Also, the EPC filter screen is modified to prevent the sides of the screen from getting sucked together, which causes low line pressure with high throttle and burns the clutch and band. You now have complete gear control to hold 1st-2nd-3rd to any rpm and you can downshift to the selected gear. Even with the TransGo shift kit, torque management should be removed and your shift points adjusted. In other words, get back on the computer.

PROJECT 25 | Installing a Skip Shift Eliminator

Time: Less than 1 hour

Tools: A small screwdriver to undo the electrical clip.

Torque: N/A

Talent: ★★

Applicable Years: 1997–2004

Tab: $100

Tip: N/A

Performance Gain: No real performance gain, just a lot less aggravation.

Complementary Work: N/A

The computer-assisted gear system (CAGS)—sometimes called the skip shift—helps the Corvette meet federal gas-mileage standards and avoid the gas-guzzler tax. The first thing you need to do is disable it. (People have been doing this for years.) The CAGS kicks in when the vehicle speed is between 15 and 19 miles per hour, the engine coolant temperature is greater than 171 degrees Fahrenheit, the barometric pressure is greater than 76 kilopascals (kPa), and the accelerator position is less than 26 percent. Once the 1–4 upshift solenoid is enabled, the solenoid will not be reenabled until the vehicle speed returns to zero miles per hour and the conditions for enabling the 1–4 upshift solenoid are met once again. In simpler terms, it's almost as if they designed it to make you go crazy driving around town.

Most people don't like the forced 1–4 shifts in a manual Corvette. The common solution is to install a CAGS disable device, or skip shift eliminator. These kits will prevent the CAGS solenoid from engaging, but the "1 to 4" light on the dash will still illuminate when skip shift conditions are met. The devices vary by manufacturer, but the installation principles are as described here.

First, disconnect your Corvette's negative battery cable from the battery. The stock cable color is black, but check the markings on the battery to be certain. Now raise the rear of the Corvette and place the car securely on jack stands.

The CAGS solenoid is located on the driver's side of the transmission toward the top and rear of the transmission. The wiring harness connector is white and blue. The connections are traditional GM connections—flip up the locking tab as you pull them apart. Gently lift the locking clip, without breaking it, and unplug the connector from the solenoid. Make sure the connector and solenoid are clean. Plug the black end of the CAGS eliminator into the wiring harness. You should hear a snap when the wiring harness is completely seated. Plug the white end of the CAGS eliminator into the CAGS solenoid. Again, you'll hear a snap when the connector is completely seated.

Now lower your Corvette back onto the ground and reconnect the negative battery cable. When you take the car for a test drive, you'll notice that the dash light still appears, and that's normal. It actually takes longer to get your Corvette up on jack stands than it does to install the disabler.

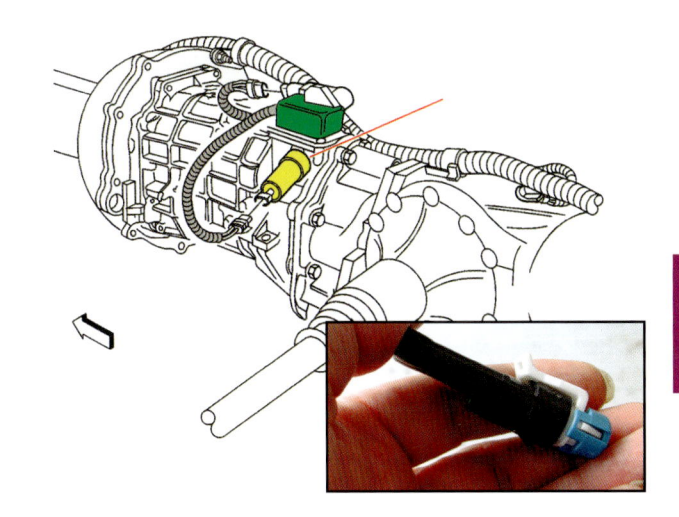

Above: The hardest part of eliminating the computer-assisted gear system (CAGS), is getting your hand up to the wiring harness. Nonetheless, I think every C5 owner with a manual transmission has managed to do it. **Inset:** This skip shift eliminator will remove all of the aggravation you experience with the CAGS skip shift system. If you purchased your C5 used, it most likely already has one in place. While you can install it rather easily, an experienced Corvette shop can put it in place in about 15 minutes.

DRIVETRAIN

PROJECT 26 | Changing the Rear Differential Lubricant

Time: 1 hour

Tools: Drain plug kit and gear oil

Talent: ★★

Applicable Years: 1997–2004

Torque: N/A

Tab: $15 to $75

Tip: Use synthetic gear oil

Performance Gain: N/A

Complementary Work: N/A

For the first time since 1962, the C5 allowed Corvette owners to change the rear differential lube in their Corvettes. GM did away with the rear differential drain plug in 1963. It finally returned in 1997. The main cause of differential bearing and seal failure is dirty lubricant, and C5 owners can easily change the rear axle lubricant in the comfort of their own garages.

Checking and Adding

Checking and adding rear differential lubricant is fairly easy, provided the last person who added oil didn't go crazy when they tightened the fill plug. It takes a special tool to loosen the drain plug, but most shops don't have the tool and resort to the standard Vise-Grip technique. If you're going to a new Corvette shop, ask how they intend to remove the fill plug. If they look at you like you're crazy, find a shop that has the right tool. Or you can buy your own specialty tool and do it yourself. Once you get the fill plug out of the case, it's simply a matter of sticking your finger in the hole and checking the level. If the level is between the hole threads and ¼ inch below the threads, the axle is full.

The factory manual calls for 75W-90 gear oil in the rear gears. This is 75W-90 synthetic oil that can cope with the extreme heat. If you use the GM oil (12378261), you'll have to add a limited-slip differential additive. Make sure that the lubricant contains a Posi-Traction additive (also called friction modifier). If you can't find the correct oil, you can purchase the additive separately. Just remember that the only thing that smells worse than gear oil is a packet of Posi-Traction additive.

What Gears Are in My Corvette?

GM used two different limited-slip rear axles in the C5 Corvette. The automatic transmission cars use a 2.73:1 rear axle ratio, while the manual transmission cars use a 3.42:1 rear axle ratio. To give the automatic folks improved acceleration, a 3.15:1 ratio was offered as an option.

Why Change the Lube?

The C5 uses a rear drive axle assembly built by Getrag in Germany. It bolts to the back of the transmission, and the transmission output shaft runs directly to the axle's pinion. The rear drive axle assembly doesn't share lubricant with the transmission. It uses a conventional, hypoid gear set with a 7 ⅝-inch-diameter ring gear and a traditional clutch-type, limited-slip differential.

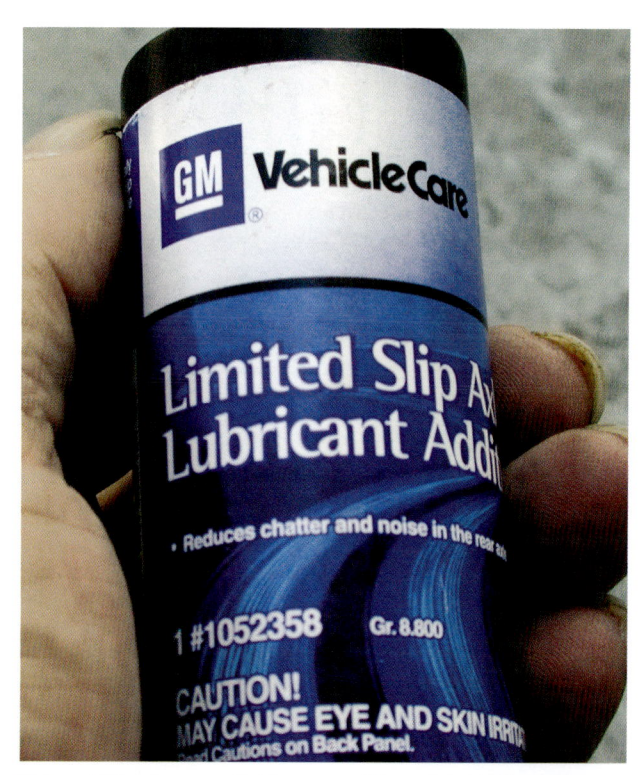

This additive is essential for your differential. Until another manufacturer demonstrates that it has a truly superior product, stick with the GM brand.

GM recommends no maintenance fluid changes, but if you run your car hard, I suggest you change your fluid every 35,000 miles. Changing the rear differential lubricant is so easy that you might even consider doing it annually. It takes less than 2 quarts of lubricant, plus 4 ounces of Posi-Traction additive. If you're going to change the fluid, remove the axle drain plug. When the fluid has drained out, replace the plug and fill the axle through the fill hole until it begins to run out. The factory-fill lubricant is a mix of GM axle lubricant (12345977), a 75W-90 petroleum-based GL5 gear oil, and GM axle lubricant friction modifier (1052358). The friction modifier eliminates limited-slip noise. The service manual recommends a mix of 6.25 percent friction modifier, or about 3.4 ounces of the axle's 1.7-quart lubricant capacity.

If you feel chatter or hear a squealing noise from the rear of the car while you make a slow-speed tight circle, it's probably the limited-slip differential. Add another ½ ounce of the modifier and road test the car again. If the noise is still present, add another ½ ounce and road test it again. Once you get to the point of no squeal or chatter, top off with a little more axle lubricant. Generally, it will take 2.5 to 4 ounces of friction modifier to solve noise problems. If you get to 4.5 ounces and the noise persists, there may be a problem with the axle, and you should take the car to the dealer for repairs.

Axle Ratios
Manual: 3.42
Automatic: 2.73 (base)
Automatic: 3.15 (optional)

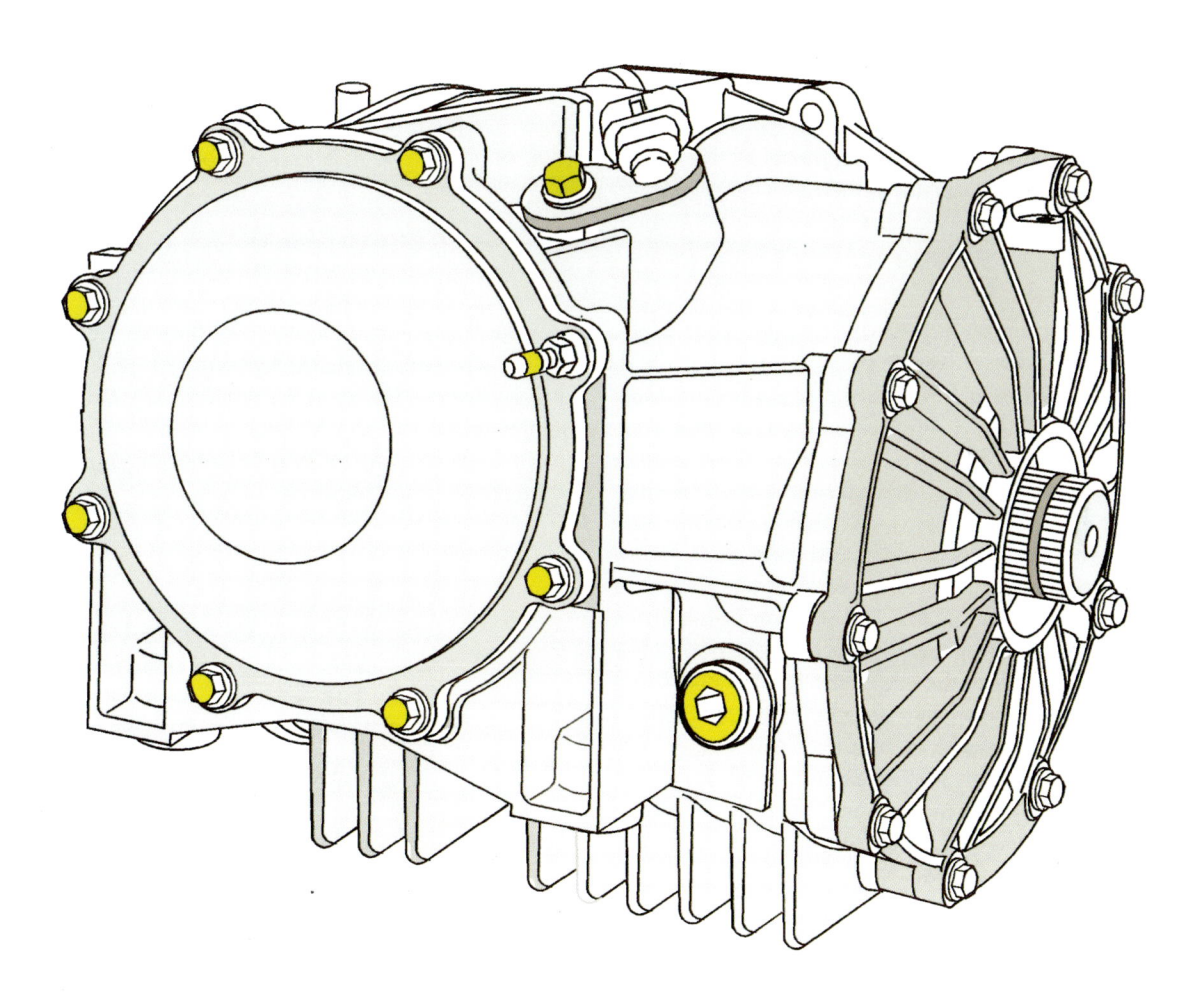

This unit is used with all Corvette engines. The transmission, whether automatic or six-speed, drives the pinion, which in turn rotates the ring gear and differential. *Courtesy of General Motors*

DRIVETRAIN

SECTION FIVE
BRAKES

Projects 27–31

| **PROJECT 27** | Bleeding and Flushing the Hydraulic System |

Time: 1 hour

Tools: Pressure bleeder, brake fluid, and vinyl tubing

Talent: ★★★

Applicable Years: 1997–2004

Torque: N/A

Tab: $10 to $100

Tip: This is a necessary task and should be done once a year.

Performance Gain: You'll improve the braking force and decrease the chance of brake pedal fade.

Complementary Work: Check your brake pads for wear. Any time you're below half of the original thickness, consider replacing them.

Very few people flush the brake system, even though you should do it once a year. If you run track events with your Corvette, you should flush the brake fluid before every event. Your brake fluid actually sucks moisture out of the air. That means after a year or two, there's a lot of moisture in your brake fluid, even if you never take the caps off the master cylinder.

When your fluid gets contaminated with a lot of moisture, you have two problems. First, the water can set up rust problems in your brake system. This problem is minimized by all the aluminum components in the C5 brake system, but the problem still exists in the steel brake lines. Second, water boils at 212 degrees Fahrenheit—a pretty common temperature for your brake system. At the track, my calipers run around 550 degrees. At these temperatures all the moisture in the brake system will turn to steam and you'll have a vapor in your brake lines, not fluid. You'll also have a very spongy brake pedal since steam is easily compressed.

Bleeding Is Not Flushing

Flushing your hydraulic system is different from simply bleeding the brake system. When you bleed the brake system, you're simply trying to get all the air bubbles out. When you flush the brake system, you're getting rid of all the fluid that's been in the system since the last flush. Actually, you're replacing the current brake fluid with brand-new, fresh brake fluid.

The net effect of bleeding and flushing are very similar. If you kept on bleeding the hydraulic system even after you had expelled all the air in the system, you would be doing a complete flush of the braking system. Remember that in bleeding we're trying to get rid of air, while in flushing we're trying to get rid of moisture.

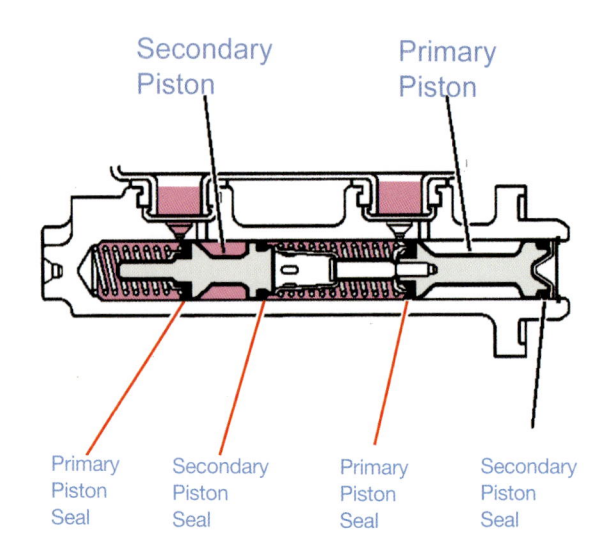

This is the C5 brake master cylinder. *Courtesy of General Motors*

Never use silicone brake fluid in a C5 Corvette. The ABS unit will aerate the fluid, and you'll end up with a mass of tiny air bubbles the first time you activate the ABS. Then the pedal will go to the floor. Instead, you can use a DOT 4 fluid with no problems.

I like to use a clear vinyl hose running from the caliper bleeder screw to a clear plastic bottle. Always place some fluid in the bottom of the bottle and submerge the hose before you start bleeding to prevent any air from getting into the system. If you use clear plastic, you can keep bleeding fluid out of the system until you see clean fluid with no air bubbles.

Four Ways to Flush, or Bleed, the Brake System

There are four methods for bleeding the brake system, but it doesn't make much difference which one you use, since they all accomplish the same thing.

Pressure Bleeding: Start with a reservoir of new brake fluid in a pressurized tank. Then place a positive air pressure force on the opposite side of the fluid, forcing it into the brake system and pushing all the old brake fluid out of the system.

Vacuum Bleeding: Fill a reservoir with clean brake fluid, and then apply a vacuum at the caliper bleeder nipple. The vacuum pump pulls the fluid through the system. The only problem with this method is that you're pulling, not pushing, brake fluid through the system—exactly the opposite way the brake fluid normally flows.

Family and Friends Bleeding: You can recruit a helper and have them push on the pedal repeatedly until the entire system is bled. This very boring job and entails a lot of complaining by the family member or friend.

Gravity Bleeding: Gravity bleeding is a unique procedure where you simply open all the bleeder screws and let the fluid run out, but never ever let the master cylinder run dry. A lot of race teams use gravity bleeding when they have completely rebuilt the brake system.

I prefer pressure bleeding the system. It usually requires a lot fewer promises for future help and, more importantly, most folks don't really know how to effectively push the brake pedal during the bleeding process.

From the Bleeder to the Bottle

It's not a good idea to simply let the fluid run all over the shop floor. The best way is to purchase a couple of feet of vinyl tubing. Push one end onto the bleeder screw and place the other end in a bottle to minimize the mess on the floor and on the car. Brake fluid is the best paint remover on the market—take care that it doesn't contact any painted surfaces. There is no such thing as being too careful with brake fluid.

Pushing the Pedal

In normal use, the piston in a master cylinder only goes into the cylinder about one-third of the distance. New master cylinders have clean and smooth pistons, but pistons on older master cylinders are dirty and even rusted or corroded.

As the piston is depressed into the bore of the master cylinder, the seals and fluid help to keep that one-third or so clean, smooth, and well-lubricated. The remaining two-thirds of the bore is exposed to dirty brake fluid and doesn't get the benefit of regular cleaning. When your helper depresses the piston during the bleeding process, the cylinder is pushed into this dirty corroded area and drags across the seals. This is a perfect situation for tearing and nicking the little O-ring seals.

The more dirt or corrosion on the interior of the master cylinder, or the more frequent and vigorous the person pumps the pedal, the worse the damage. The net effect will be a leaky master cylinder. The best way to prevent this is to place a block of wood under the brake pedal so the helper in the seat can never push the brake pedal all the way to the floor. Also, ask your helper to pump the pedal with slow and even strokes.

The Sequence

There's a sequence that works best when you're bleeding or flushing the brake hydraulic system. Always follow the sequence for your model year.

1997–2000: Right rear, left rear, right front, left front
2001–2004: Right rear, left front, left rear, right front

PROJECT 28 | Replacing the Brake Pads and Rotors

Time: 1 hour

Tools: Metric box wrenches, ½-inch metric sockets, and brake piston tool.

Talent: ★★★   **Applicable Years:** 1997–2004

Torque: N/A

Tab: $75

Tip: Purchase the special tool that pushes the caliper piston back into the caliper. The tool is much better than using a pair of pliers, and it's very inexpensive.

Performance Gain: New pads and rotors will allow for greater braking power.

Complementary Work: Consider installing new hydraulic brake hoses at this time.

Doing a brake job at home is simply a matter of checking everything carefully and taking your time. Take one side apart at a time. That way, if you get confused, you can use the other side for a reference.

The C5 Corvettes used the same brake system for their entire production run. This brake system is fine for the street-driven Corvette. Just remember to use a top-quality brake pad designed for street driving. The previous owner may have skimped on the maintenance, so check everything carefully.

Before you start taking your Corvette apart, ask yourself what brand of brake pad you want to purchase. This means you have to decide what's important to you in a brake pad. If you only drive your Corvette on the street, your biggest concerns will be about brake dust and noise.

Almost every brake pad on the market will stop your street-driven Corvette. Most of the major brands also come with a lifetime warranty. This is another case where you need to ask questions. Find out what brake pads other Corvette owners like.

The really important thing when you change brakes pads isn't the composition of the pad but the preparation of the rotor. Because stopping involves the brake pad rubbing against the rotor, the rotor and the brake pad each represent one-half of the equation.

It's also critical to break in your new pads carefully. Take the car for a drive and get it up to about 40 miles per hour. Then bring the car to a complete stop. Drive a mile or so and

Brembo brakes are the world standard for brakes. They're not cheap, but they hold up better than any other rotor available.

Once rotors get below a certain thickness, they simply can't dissipate the heat properly. Corvette C5 front brake rotors begin life at 32 millimeters (26 millimeters for the rear), and they need to be thrown out when they get below 30.3 millimeters (24.5 millimeters for the rear rotors). Take measurements at several different places on the rotor and compare the numbers to check for thickness variation. There should be no more than 0.013 millimeter (0.0005 inch) variation between your measurements.

This is one very impressive caliper and rotor. Even though this example is on a racing car, you could also use it on the street.

Make sure you don't lose this little clip when you replace the brake pads. If you purchase your brake pads from the Chevrolet dealer, they generally include a new clip. Most of the aftermarket brake manufacturers do not include it, however, so you'll need to reuse the old ones.

repeat the process. Do this about 15 times and you will have started the bedding process properly. Just be careful for the next 100 miles and you'll have a great set of brakes.

Brake Rotors

Warped brake rotors are not as common as people think. The brake pedal may feel like it's pushing back at your foot, but warping is not always the reason. When the new brake pads were installed, the old rotors may have been left on the car. These rotors would be coated with the friction material from the previous pads. When you apply the new brake pads, they never even touch the rotor; they simply interact with the old friction material. The brake pads will grab and the whole braking application will feel like the rotors are warped.

In general, though, the C5 Corvette seems to have brake rotor issues. At least we're starting to come across a lot of warped brake rotors. The stock rotors work just fine for the average person who drives on the street. When it becomes time to replace the pads and rotors, however, too many people start price shopping. Cheap rotors simply won't hold up in the C5.

Whether it's the metallurgy of the cheap aftermarket rotors or the design of the internal vanes, most aftermarket rotors don't work well on the C5. Even if you get a good deal up front, consider the price of replacing the rotors after only a few thousand miles. It's worth it to pay for the OEM rotors.

Brake Squeal

Squealing brakes are the single biggest complaint that repair shops have, but brake squeal is generally not a safety hazard. Brake squeal is really a vibration that can be prevented if you spend an extra half-hour installing your new brake pads.

When the pad touches the brake rotor, it sets up a vibration, or oscillation. The noise you hear when you pull up to the red light is really a vibrating rotor or caliper. The noise you hear when you strike a tuning fork is nothing more than a vibrating piece of steel, and the same idea applies to your C5's brake rotors and calipers.

When you install new pads, you not only need to measure them for thickness and runout, you also need to prepare the surface properly. With an orbital sander and 100-grit paper, sand both the inside and outside surfaces of the rotors. You want a series of small circular sanding marks on the rotor.

Once you have them sanded, scrub your rotors in soap and hot water to remove all the little metallic particles. Once you've done a good job of scrubbing the rotors, dry them off and spray them with brake cleaner. When you're all done with this, take a white paper towel and run it across the surface. It the towel shows dirt then you have some more cleaning to do.

Unfortunately, once you have a squeal, it's very hard to get rid of. It is better to make sure the brake pads are installed with a lot of care than to try to stop a noise once it starts. I've installed several hundred brake pads and only a few ever developed a squeal. Over the long haul, it's better to be right the first time—especially when it comes to stopping.

BRAKES

PROJECT 29 | Installing Stainless-Steel Brake Hoses

Time: 2 to 3 hours

Tools: N/A

Torque: N/A

Talent: ★★

Applicable Years: 1997–2004

Tab: $15 to $145

Tip: Make sure you completely bleed the braking system once you have the new lines installed.

Performance Gain: If your car has over 40,000 miles, you'll notice a much firmer brake pedal.

Complementary Work: This is a good time to flush and replace all the brake fluid in the system.

If your C5 has less than 50,000 miles and is only a few years old, installing stainless-steel brake hoses probably doesn't make much sense. But if you have a lot of miles and your C5 is one of the earlier cars, this could be an improvement.

The braking system is designed to operate on pressure. When you push on the brake pedal, you move a rod and piston into the bore of the master cylinder. The size of the bore and the ratio of the force applied have all been carefully calculated by the good folks at General Motors. As a result of the movement, the brake fluid is pressurized so that it can move the caliper piston against the brake pad, which in turn pushes the pad against the rotor in an effort to create heat. Remember, the creation of heat is what actually stops your car. The pure physics of the braking system is that we're converting mechanical energy into thermal energy. Remember, energy cannot be destroyed, it can only be converted. The entire braking system on your Corvette is really just a giant energy conversion system. The more mechanical energy you can convert to thermal energy, and the faster you can do it, the quicker your Corvette is going to stop. Why do you think the NASCAR teams are running 1500 degrees at Richmond twice a year?

Everything on your brake hydraulic system is steel, except for the rubber brake lines at the four corners of the car. These are the only places where hydraulic pressure can create a problem, and it really doesn't become a problem until the car ages. Eventually these rubber brake hoses will start to deteriorate so that when you push down on the brake pedal, the rubber hose will actually expand. This means you have a lag before the pressure acts on the caliper piston. You may notice a slightly softer pedal than the car had when it was brand-new.

Brake lines have traditionally been made from rubber tubing with steel or aluminum connectors crimped onto the ends. Nearly all passenger cars are shipped with rubber brake lines, and they hardly ever fail. Stainless-steel brake lines are made with Teflon tubing, not rubber. Teflon has a number of

Most people install these stainless-steel braided brake lines for their looks. They also cost a whole lot less than the original equipment flexible brake lines. You only need these if the car is old and the original brake lines are starting to deteriorate. But be forewarned that they have a tendency to break right at this point.

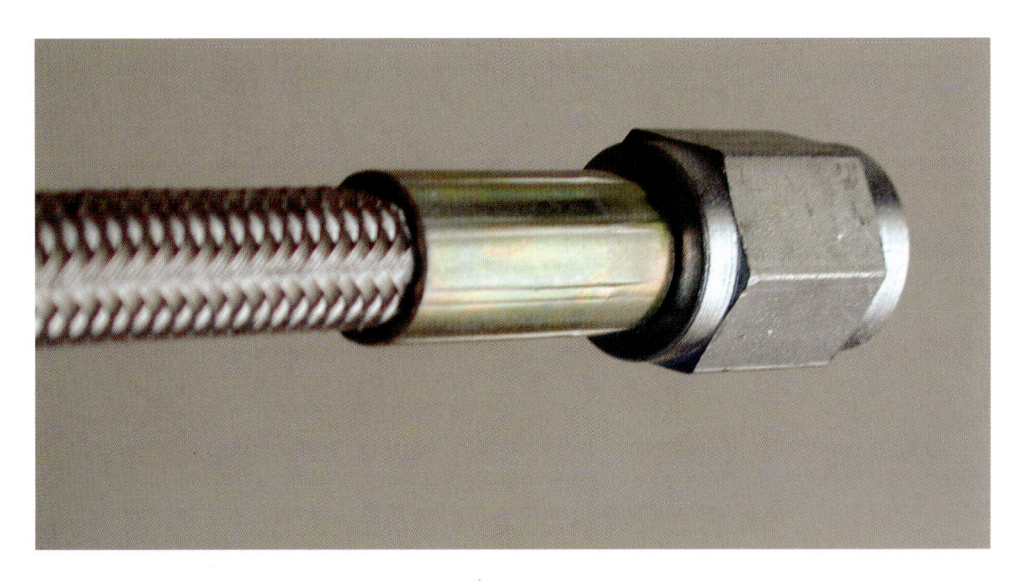

This massive six-piston Wilwood caliper may very well be the best caliper you can install on your C5, but it does require a lot of adapters for the brake line installation. Any time you install new brake lines, you increase the potential for leaks. Check every fitting carefully before you actually drive your Corvette.

advantages over rubber, chiefly that it doesn't expand under pressure or deteriorate with age. Teflon also resists high temperatures and is chemically inert, so it's compatible with all brake fluids.

Teflon is fragile, so it has to be protected from physical damage (chafing, flying rocks, etc.). Although some manufacturers armor their Teflon hoses with Kevlar, most protect the Teflon with an external sheath of braided stainless-steel wire.

GM uses rubber brake lines on the Corvette because they work. There's very little to be gained by replacing the rubber lines on a car that's less than five years old.

That's why armored Teflon hose is usually called stainless-steel hose. Stainless-steel brake lines are really Teflon lines with a protective stainless-steel-braid cover.

The ends of the hoses have to be securely attached to the brake calipers and the hard lines, so each hose is terminated by threaded hose ends. The cheap way to attach the hose-end fittings to the hoses is to crimp or swage them onto the hoses, like the fittings on rubber hoses. The more expensive way is to use a two-piece replaceable hose end that captures a portion of the hose between an inner nipple and a concentric outer socket.

The Problems

There are some big downsides to the stainless-steel braided brake hoses. First, the Teflon inner liner in the stainless-steel brake lines tends to take a "cold set" after a bit, and any bending movement beyond a limited degree will cause the inner liner to kink.

Second, until a few years ago the braided lines broke right at the fitting during prolonged use. Aftermarket stainless-steel brake lines break a lot easier than the stock GM hoses, and a lot of them fail to pass the Department of Transportation tests. Make sure that you only purchase braided lines that have met the DOT specifications (DOT FMVSS106). Because of these problems, stainless-steel braided hoses need to be checked regularly.

Installation

The secret to installing a braided brake hose is to work quickly. Remove the old hose and get the new one in place before the master brake cylinder runs dry. As you replace each of the brake lines, go back up to the master cylinder and fill the reservoir. You don't want to let the system run dry during this installation.

PROJECT 30 | Changing the Brake Bias

Time: 2 to 3 hours

Tools: 19-millimeter combination wrench, 13-millimeter combination wrench

Torque: N/A

Talent: ★★

Applicable Years: 1997–2004

Tab: $15 to $145

Tip: Make sure you completely bleed the braking system once you have the new bias spring installed.

Performance Gain: You'll stop a little faster and with less nose dive on dry pavement. Just be very careful in wet weather since the rear brakes are going to have more force.

Complementary Work: This is a good time to flush and replace all the brake fluid in the system.

Some folks want to add more rear bias to their brake system, so they install this handy spring sold by Doug Rippie. Here's how to do it:

(1) You'll need to bleed the brakes after you replace the spring, so you might as well jack up the car from the very first step and remove all the wheels.

(2) Locate the brake proportioning valve junction box. It's located below the forward section of the master cylinder. Mark the junction box where the input pipe from the master cylinder enters. Put an old towel on the frame underneath to catch any brake fluid that might leak out during this process.

(3) Use a 13-millimeter wrench to remove the brake pipes from the master cylinder to the junction box, and to remove the pipe from the junction box toward the front brakes. You may need to put a piece of tape over the pipe from the master cylinder to keep fluid from leaking (1 tablespoon or so of fluid).

(4) Wrap the junction box in a rag and put it in a vise with the 19-millimeter retaining nut pointed up. Carefully remove the nut. There's a fiber retainer on the end of the nut. Don't misplace this fiber retainer. The old spring should be protruding about 1 inch.

(5) Remove the old spring. If the valve comes out with the spring, you'll need to reinstall the valve on the new spring in the same orientation.

(6) Insert the new spring/valve in the junction box.

(7) Here's the hard part: the new spring is considerably stiffer than the stock spring. You need to put the fiber retainer on the end of the nut and then place the nut assembly

The front of the C5 master cylinder is sealed. That means that you'll have to remove the assembly from the car to install the new bias spring. It helps if you remove the brake fluid from the plastic reservoir before you start on the job. A turkey baster works nicely for this task.

Make sure you use a flare nut wrench to unhook these lines. Also, put a rag or towel under the assembly to catch any dripping brake fluid. Brake fluid will remove paint better than any paint remover you might find in the store.

One way to check brake bias at the track is to use a pyrometer for measuring rotor temperature. In an ideal world, the front and rear rotors should be the same temperature. That indicates that both ends of the car are working at an optimum level. In a stock C5 the front brakes do most of the stopping, which is done for safety and legal reasons. A track car, or a race car, is totally different. You want both ends of the car at an optimum level of braking.

over the end of the spring. Now you'll have to compress the spring to get the nut to the threads. I've used a small, flat bar to push down on the nut while turning it with the wrench. I didn't find a torque specification for the nut, but it shouldn't be much.

(8) Reattach the proportioning valve assembly to the brake pipes. Attach the pipe from the master cylinder to the fitting you marked in step 2. Tighten the brake pipes to 13 ft-lb.

(9) Bleed the brakes. (For methods, see Project 27.) Make sure you use the proper sequence for bleeding. Replace the wheels, and lower the car back to the ground.

(10) Before you start the car, pump the brakes. After a couple of pumps, the pedal should sit really high. Start the car and the pedal should drop back to normal position. Check the brake fluid level and top off if necessary.

(11) Make sure the brakes work before you try any high-speed or emergency stops. Drive back and forth in the driveway until you're comfortable that the brakes work. If the pedal is spongy or drops to the floor, then you need to bleed the brakes again.

(12) If the driveway test works, take your C5 for a test drive. Start out doing complete stops from 30 miles per hour. Please don't do any panic stops until you're sure that everything is safe.

(13) Check the fluid level again in a couple of days, just to make sure you don't have any leaks.

With more rear brake bias you could encounter some problems in wet conditions. The ABS system should take care of any tendency to lock up the rear wheels, but there's always a chance. Wait for a really nice rainy day and try out the system in a large, wet vacant parking lot.

Here are the two different springs. The stock spring is the lower one. The hardest part of this whole job is compressing the new spring into the master cylinder. The strength of the spring determines the front/rear bias in the C5 braking system. When you change the bias spring just keep in mind that you've increased the possibility that the rear brakes will lock up first in wet weather.

BRAKES

PROJECT 31 | Replacing an ABS Sensor

Time: 1 hour

Tools: Metric sockets (½-inch drive)   **Torque:** N/A

Talent: ★★   **Applicable Years:** 1997–2004   **Tab:** $50

Tip: Keep everything as clean as possible during the wheel sensor replacement.

Performance Gain: Your anti-lock braking system (ABS) will work properly.

Complementary Work: This is really a stand-alone project, and you don't need to delve into the entire braking system.

The hardest part of this project is finding out which one of the four ABS sensors has failed. Let a professional diagnose your ABS problem since the scan equipment needed to perform a complete diagnosis is simply not practical for the weekend warrior.

Arrange to have a good Corvette shop diagnose the reason your ABS light is coming on. If it's only the wheel sensor, you can drive the car home and replace the appropriate wheel sensor yourself. You aren't doing this to save money; rather, you're doing it because you like working on your car. If you're a regular customer at a Corvette shop, they'll understand this and work with you accordingly.

Remember that an ABS failure is not a brake system failure. ABS systems are designed in such a way that when your ABS system fails you still have a normal brake system.

Very few of us even get to the point where we activate the ABS system during normal driving. I suggest that on a nice, wet Saturday morning you find a totally deserted area of road or parking lot and brake hard enough to activate your ABS system. Learn what the brake pedal does under heavy braking and how easy it is to control the car. You also need to learn how to keep your foot on the brake pedal even under conditions that activate the ABS. Most of us grew up with the notion that when your brakes start to lock up you should come off the brake pedal slightly to regain control. But with ABS you can just keep standing on the brakes until the car comes to a complete stop. The ABS system will pulsate the brakes for you. Even though you understand this, you need to actually experience the operation of the ABS brake system.

The Wheel Sensor

Let's assume that you, or a service facility, have determined that one of your ABS wheel sensors is defective. The replacement couldn't be easier. First, trace the sensor wires back to the connector and unplug the sensor from the harness. Go back to the sensor and remove the two small bolts that hold the sensor in place. You also have to remove the brackets and grommets that are used for routing the wiring. This sensor won't just pull out

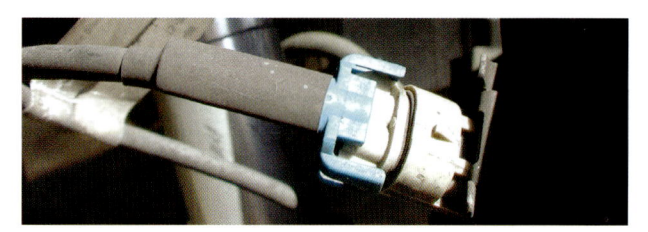

It's very easy to unplug the sensor from the ABS harness. No tools are required. The most difficult part of working with the ABS sensors is getting the diagnosis correct, because problems are so rare. Have the diagnosis done by a professional and then replace the sensor in your home garage.

when it's been there for a while, so you may have to help it. Fortunately, you've already determined it's defective, so you don't have to worry about damaging it. On the other hand, you need to be very concerned that you don't damage the aluminum suspension knuckle. Pry bars and screwdrivers may help you get the sensor out of the hub, but they could also scratch the hub.

Once you have the sensor out, clean the area well. Brake cleaner works best in this situation, but the silicone will fight you a little bit. Just keep telling yourself that the cleaner you get this area the greater your chances for success. You really don't want to go back to the shop that did the diagnosis and admit that you couldn't properly install the sensor.

The new sensor will have a coating since it's going into the aluminum. The Corvette engineering team spent a lot of time working with these coatings since the Corvette was one of the first cars to use aluminum suspension pieces held together with steel bolts. The ball joints and ABS sensors all have special coatings designed to ward off any possible corrosion problem. Once you're happy that everything is cleanl, it's simply a matter of putting the new sensor into its hole. It helps if you twist a little as you push it in place. Then bolt it back to the hub and connect the sensor to the wiring harness.

SECTION SIX
SUSPENSION

Projects 32–38

PROJECT 32 | Lowering Your C5

Time: Several hours

Tools: Floor jack, jack stands (2), wheel blocks, needle-nose pliers, hacksaw or serrated knife, 10-millimeter ratcheting wrench or 10-millimeter socket wrench with swivel extension, 18-millimeter combination wrench or 18-millimeter socket wrench, and 13/16-inch combination wrench or 13/16-inch deep socket wrench

Torque: N/A

Talent: ★★   **Applicable Years:** 1997–2004

Tab: No real expense, except for the post alignment

Tip: You only want to lower the car, not slam it.

Performance Gain: Any time you lower your Corvette it should handle better. The exception is if you get carried away and lower it so much that you end up hitting the suspension bump stops.

Complementary Work: This could be a good time to replace your stock sway bars with larger diameter bars to increase the cornering ability.

Lowering your Corvette's center of gravity will not only make your C5 look better, it will corner better as well. If you go too low, you'll run into some problems. The shocks could easily bottom out or the car could end up riding on the bump stops. When that happens, you essentially have no suspension and the car will handle poorly.

You need to be fastidious about ride height when you start to lower your C5. Ride height is critical on the C5. You also want to retain the slight rake that GM designed into the car.

Lowering the Rear
(1) Place blocks or boards under the front wheels to prevent the car from rolling.
(2) Using the floor jack, raise the rear of the car, ensuring that the jack is positioned in the center of the rear crossmember.
(3) Place two jack stands under the rear crossmember, and slowly lower the car onto them. Make sure you've raised the car high enough to allow plenty of work space.
(4) Locate the rear leaf spring where there's a bolt on each end (18-millimeter bottom, 13/16-inch nut on top) with a rubber bushing on top and bottom.
(5) At the top of the bolt above the nut, you will see a small C-clip. This clip must be removed in order to get the nut off the top. Use needle-nose pliers to remove the clip. Only remove the bolt completely if you wish to lower the car more than 1 inch in the rear. It's possible, in most cases, to lower about 1 inch without removing the bolt at all and without cutting the bushing. Cutting the bushing will make your ride a lot harsher than before.
(6) Remove the bolt and bushings from the leaf spring (the leaf spring isn't under pressure).
(7) Cut both bushings so that the section with the extension is the only one left. Use a hacksaw or serrated knife and remove only a little at a time.
(8) Once the bushings are cut, reassemble the bolt, bushings, and nut through the leaf spring and A-arm. Make sure that you leave at least two threads showing from the top of the nut when you retighten. It'll seem loose, but when the car is lowered onto the ground, the weight will load the spring. Don't forget to replace the C-clip after you tighten the nut.
(9) Lower the car, and the rear is done.

97

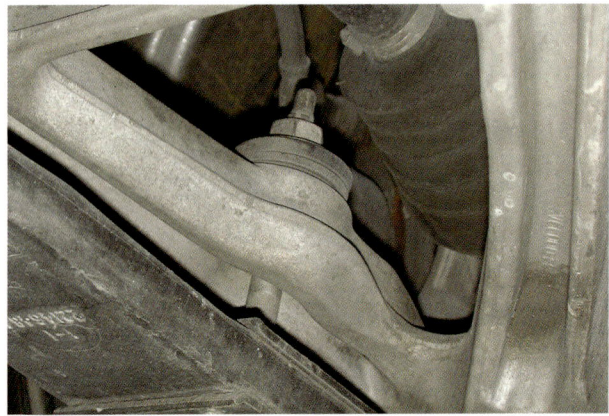

Here's the left rear corner on a C5. At the top of the bolt, you'll find a clip. Once you remove the clip, you can simply turn the bolt to lower or raise the car, but don't try to lower it more than 1 inch.

The bolt for changing the ride height isn't easy to display in a picture, but it's easy to find if you're working on the car. Change the ride height one turn at a time. Then drive the car around the block and take a new set of measurements.

Lowering the Front
(1) Make sure the car is in gear to prevent it from rolling.
(2) Raise the car with the jack positioned in the center of the front crossmember. Loosen the lug nuts before you get the wheels off the ground.
(3) Place the two jack stands under the front crossmember, and slowly lower the car onto them.
(4) Remove the front wheels.
(5) Locate the end of the front leaf spring and, with a 10-millimeter wrench or socket, turn the bolt as if you were loosening it. (These bolts use left-hand threads so you are actually tightening them when you turn them in a counterclockwise direction.)
(6) Turn the bolt until there's no longer a gap between the bushing and the bolt. Measure the exposed threads to ensure both sides are done the same. On a Z51, the car will not drop as much as one with the standard suspension.
(7) You can completely remove the front lowering bolt if you would like to go even lower, but you won't have any adjustability left for setting the ride height. That's the main reason for leaving the bolts in their location. Use a spreading tool or remove the A-arm in order to totally remove the adjusting bolt and bushings. Remember, that's not something I recommend.
(8) Put the wheels back on and lower the car to the ground. You may need to place a board under the tires to be able to remove the floor jack once the car is lowered.

Now that you've gotten everything completed, take the car for a 20-mile ride to get everything settled into position. Next, you're going to make some very careful measurements on a very level floor. You want the car to sit level from right to left. If you're really fanatical, you can place yourself in the driver's seat while a friend measures the ride height.

I suggest that you only lower the car to about half of what you really want. It's so easy to change the stance of the C5 that you don't need to do it all at once. Bring the car down in stages. That way, if you encounter a problem you know that a slight increase in ride height will solve the problem. Lower it a little, drive it for a while, and then decide if you really want to go lower.

This is a race car with an aftermarket bolt.

SUSPENSION

98

PROJECT 33 | Installing Polyurethane Bushings

Time: A couple hours to several days

Tools: 18-millimeter socket and 18-millimeter combination wrench, T40 Torx driver

Torque: Front bracket: 58 N-m (43 ft-lb); rear bracket: 65 N-m (49 ft-lb); nuts: 95 N-m (70 ft-lb); sway bar link nuts: 72 N-m (53 ft-lb)

Talent: ★★ for the sway bars, and at least ★★★ for the control arms

Applicable Years: 1997–2004   **Tab:** $50 to $500

Tip: Start with the sway bar bushings and then do the control arms.

Performance Gain: The handling will be greatly improved, although you may only notice it on a racetrack. The stock Corvette is that good. On the other hand, if your C5 has over 100,000 miles, you may notice a huge difference.

Complementary Work: This is a good time to swap your stock sway bars for larger-diameter bars. Since you're doing this to improve the handling of your C5 you might consider new shock absorbers as well.

Very few C5 Corvettes have worn-out sway bar bushings at this point, but the time is coming. The purpose of any bushing is to control motion and still allow movement. As the bushings wear, they allow more play and this play is usually inconsistent. The deteriorated rubber in the bushing must first compress before any suspension movement actually takes place. It's as if your suspension has to move a certain distance before the suspension actually comes into play. Polyurethane bushings reduce this amount of movement to almost nothing. The result is a suspension that will feel more precise and more predictable.

Sway Bar Bushings

The sway bar bushings are really easy to deal with. The real question is what to do about the original sway bar. You might consider replacing the current one with a larger version. Jump ahead a little and read Project 35 on choosing a sway bar. It's best to make your choice before you order the polyurethane bushings since each sway bar will require a different-size bushing.

This project assumes that you have a 30-millimeter solid sway bar up front. If you can't put together a 30-millimeter front and 24-millimeter (or 26-millimeter) rear bar combination, then leave everything alone. Just replace the bushings and start looking in salvage yards for the correct combination.

The rear sway bar end links for the 1997–2004 Corvettes come in three different flavors. If you have an early car with the plastic end links, this is a good time to swap them out for the aluminum end links.

The bushings that hold the sway bar to the chassis are usually in pretty good shape. They don't do much except hold the bar in place. But water gets retained in the area between the bushing and the sway bar, creating rust. This rust builds up and keeps the sway bar from rotating, meaning that your sway bar actually starts acting like an additional spring and adds a lot of stiffness to the rear suspension.

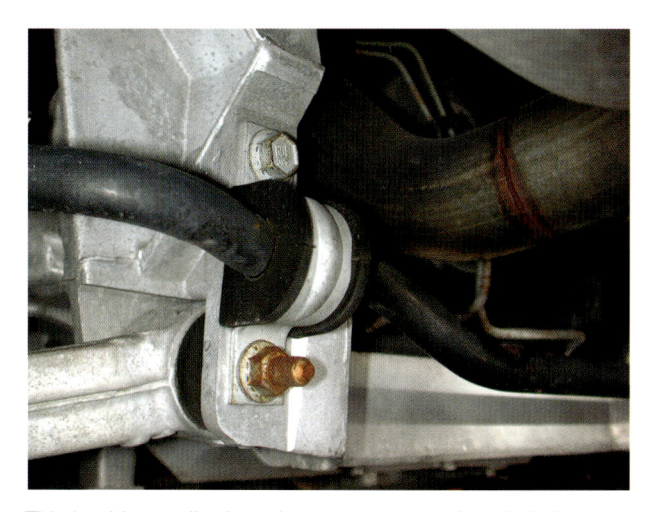

This bushing really doesn't get compressed a whole lot. The sway bar simply rotates in this bushing. You can spray some rubber lubricant into the area where the bar comes into contact with the bushing. For even better lubrication, remove the bracket, slide the bushing aside, and then lubricate the area.

SUSPENSION

Polyurethane bushings can be used to firm up suspension response in cornering maneuvers. When turning a corner, force is applied from the lower control arm to the chassis. With a worn-out bushing in this location, some compression of the bushing would take place, reducing response time. Polyurethane bushings reduce the amount of compression, which in turn sharpens the cornering forces. In other words, you can go a little faster through the corners.

Remove the sway bar and check the area carefully where the bar rides in the rubber mounts. You may need to sand rust off the sway bar. Then coat the inside of the bushing with chassis grease before you install the bushing onto the bar. Before you attach the links, check to see that the bar moves freely. You should be able to move it with one hand.

I usually go one step further and install grease fittings in the sway bar bushing mount to help prevent rust buildup between the bar and the mount. In an ideal world, you should be able to move the bar through its rotation with very little effort. New polyurethane bushings and grease will allow that to happen.

The bushings are rather inexpensive. Installing a set of bushings into the control arms can get expensive, though, if you have to pay someone to do the complete job. But doing the job in your home garage isn't easy either. Remember that all of the control arms have to be removed in order to install the bushings.

The control arm bushings last an incredibly long time and work very well on a street car. As the C5 ages, you'll see cracks start to form in the bushings. The cracks are the first indications that you should consider polyurethane replacements.

PROJECT 34 | Replacing the Shock Absorbers

Time: 2 to 3 hours

Tools: ⅜-inch drive metric sockets   **Torque:** N/A

Talent: ★★   **Applicable Years:** 1997–2004   **Tab:** $300 to $2,500

Tip: Ask other owners which shocks they have. Manufacturers are very stingy about providing any real data on their shock absorbers.

Performance Gain: New shock absorbers may help with the wheel hop problem that exists under hard acceleration.

Complementary Work: This is a good time to replace the stock sway bar bushings with polyurethane bushings.

The C5 was designed for a comfortable ride. The early C4 was designed to corner faster than any other car in the world. Between 1984 and 1996, GM spent a lot of time, and money, trying to get some ride quality into the C4. They weren't about to make the same mistake with the C5 Corvette. Ride quality was paramount this time.

The idea with the C5 was to provide decent handling but not at the expense of ride quality. You can easily improve the handling of the C5, but you'll be changing the quality of the ride at the same time. If you're used to driving a European car, this will be no problem. But if your other car is a Lexus, you'll notice a big difference. Remember that the Lexus was the ride and handling target for the C5. Interestingly, when it came to ride and handling parameters for the C6 (or 5.5) BMW was the target.

Shock absorbers are really easy to change. The hard part is deciding which ones to use as replacements since the numbers that allow you to make a decent comparison are hard to come by. GM probably spent more money engineering the stock shocks on the C5 than all of the aftermarket companies combined. If you're going to change the shocks on your C5, you should have a strong reason for doing so. I changed the shocks on my car for track use, greatly enhancing the ability of the tires to stay planted under braking and acceleration (massive rebound), but they made the car painful on the street.

Remember that shock absorbers work in two directions. I prefer a shock with a lot of rebound to keep the tires planted on the ground, but there is almost no ride quality with this type of shock. The rebound force is the force that extends the shock, or the force that keeps the tires pushed into the ground. Compression is the force that's needed to compress the shock.

In addition to these two forces, each shock absorber operates differently at high shaft speeds and low shaft speeds.

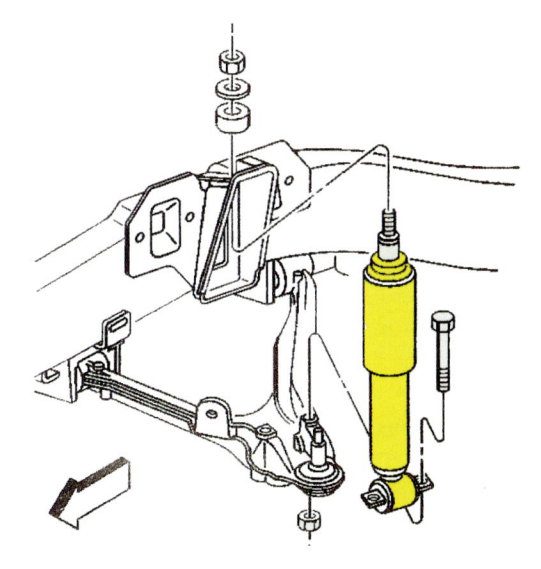

It is difficult to compress the front shock enough for removal. I find pushing up on the bottom works best. Don't remove the lower control arm, as some people have suggested; that's just unnecessary. *Courtesy of General Motors*

This is sometimes confusing because shock absorber speed has very little to do with vehicle speed. If you're going 100 miles per hour down a smooth interstate, the shaft in the shock absorber moves at a rather slow speed. When you go 40 miles per hour down a street with undulating pavement and potholes, the shaft moves rapidly. Thus, we may need different rebound and compression forces at different shaft speeds. This is why it's so hard to compare and decide among the variety of shocks that are offered for the C5 Corvette. Unless you look at the shock specification sheets (which shock manufacturers are reluctant to share with you), you're

SUSPENSION

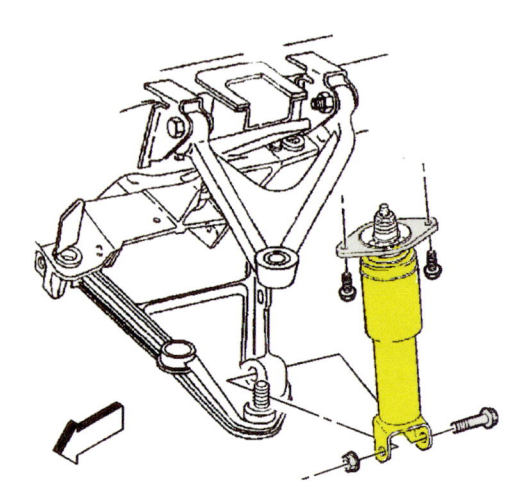

On the rear shock, it is difficult to get the lower mounting holes aligned. I use an alignment pin that looks like a giant center punch. *Courtesy of General Motors*

just guessing in the dark and trying to understand useless advertising claims. We're back to the old "ask other Corvette owners what they've used" technique.

Front Shock Removal

The front is difficult because you have to compress the shock in order to remove it. Using a jack on the lower control arm helps a little. Some people have actually removed the control arm, but there's no need to do that.

(1) Raise and support the vehicle.
(2) Remove the tire and wheel from the front.
(3) Disconnect the real-time damping electrical connector from the shock if you have the optional suspension.
(4) Move the windshield washer reservoir and the coolant overflow reservoir out of the way so that you can access the shaft on the Bilstein shock with a Torx bit while you use an open-ended wrench to turn the nut. The only way to get the Torx bit into the top of the shock is through a small hole, which becomes obvious once the reservoir bottle is moved out of the way. This is pretty easy and takes only a minute.
(5) Remove the upper mounting nut, insulator retainer, and insulator.
(6) Now turn to the lower mounting bolts, and remove all the bolts and nuts.
(7) Next comes the tricky part. You're going to have to compress the shock in order to remove it from the upper shock tower and the vehicle. I generally use a pry bar that's wrapped in a shop rag to prevent scratches.
(8) Remove the insulator and retainer from the shock absorber.

Front Installation

(1) Install the new retainer and the lower insulator onto the shock absorber.

(2) Slide the shock absorber into the upper shock tower.
(3) Install the upper insulator, retainer, and nut. Tighten the shock absorber upper mounting nut to 26 N-m (19 ft-lb).
(4) Install the shock absorber lower mounting bolts and nuts. Use a floor jack to lift the entire suspension assembly up until the bottom mounting holes on the Bilstein almost line up with the holes in the lower A-arm. The Bilstein mount will be at the wrong angle, causing the bolts to not fit correctly. I use a long alignment pin inserted into one hole and rotate the mounting plate until the holes line up. Once the first bolt is in, it's much easier to get the second in place. Then tighten the nut on the bottom and you'll be all set. Tighten the lower shock absorber mounting nuts to 28 N-m (21 ft-lb).
(5) Connect the real-time damping electrical connector to the shock if one was removed.
(6) Install the tire and wheel assembly.
(7) Lower the vehicle.

Rear Removal Procedure

Some people remove the tie rod for more room. I prefer to compress the shock and wiggle it out.

(1) Raise and support the vehicle.
(2) Remove the tire and wheel assembly.
(3) Disconnect the rear shock absorber solenoid electrical connector if your Corvette has one.
(4) Remove the lower mounting bolt that fastens the shock absorber to the lower control arm.
(5) Remove the shock's upper mounting bolts.
(6) Remove the shock absorber from the lower control arm and shock tower.
(7) Remove the upper insulator retainer and insulator from the shock absorber. Transfer the plate over to the new shock.

Rear Installation Procedure

(1) Install the upper insulator and insulator retainer (the plate) onto the new shock absorber.
(2) Bolt the shock absorber to the shock tower and lower control arm.
(3) Some people use two floor jacks—one to lift the rear of the car and another to lift the entire A-arm to install the top mounting plate bolts.
(4) Install the upper shock absorber mounting bolts, doing the rear bolt first since it's easier to line up. Don't tighten it until the second bolt is in place. Once it's fairly tight, the plate itself will be closer to the other hole, making it much easier to line up. Torque both bolts.
(5) Install the lower shock absorber mounting bolt. Tighten the upper mounting bolts to 29.5 N-m (22 ft-lb).
(6) Tighten the shock absorber lower mounting bolt to 220 N-m (162 ft-lb).
(7) Connect the shock absorber solenoid electrical connector, if equipped.
(8) Lower the vehicle.

PROJECT 35 | Choosing a Sway Bar

Time: 2 to 3 hours

Tools: 18-millimeter socket and 18-millimeter combination wrench, T40 Torx driver

Torque: Front bracket: 58 N-m (43 ft-lb); rear bracket: 65 N-m (49 ft-lb); nuts: 95 N-m (70 ft-lb); sway bar link nuts: 72 N-m (53 ft-lb)

Talent: ★★   **Applicable Years:** 1997–2004   **Tab:** $400 to $600

Tip: Mark the top of the bars to prevent installing them upside down.

Performance Gain: Greatly improved handling with minimal loss of ride quality.

Complementary Work: This would be a good time to change the shock absorbers.

I love big sway bars. The bigger they are the more I like them. Sway bars are the best way to control body roll in a corner. The idea is to connect the right and left sides of the car together with what is essentially a large torsion bar. As the car goes around a corner and the inside of the car begins to rise, this torsion bar, or sway bar, pushes the inside of the car back to the ground. Thus the car corners flatter and faster. Sway bars do not affect ride quality because they only come into play when the car is turning.

End Links
The C5 sway bar is connected to the lower suspension arms with an end link.

The standard factory links work just fine, especially if you have the aluminum end links. In 1997, GM installed plastic end links on all of the early C5 Corvettes. A lot of people get concerned when they discover that the early C5 sway bar links are plastic, but it is no big deal. I've never heard of one breaking in normal use.

If you run track events with your C5, it's not a bad idea to replace them with the Z06 aluminum links. Actually, GM may have already done that for you. Once the Z06 was introduced with aluminum sway bar end links, the entire Corvette line was switched over. This ended a noble experiment in high-tech plastic technology.

The next step up is to install adjustable links. These use Heim joints, which work nicely on race cars, but aren't a great idea for a street-driven Corvette. They're noisy and need to be cleaned on a regular basis.

You may want to remove the paint from the sway bar at the insulator contact area and lubricate the inner diameter of the isolator with synthetic grease. This will prevent any rust buildup in the area under the bushing.

If you don't have the aluminum sway bar link, you should stop at your local Chevrolet dealer and order a new set. Installation is so easy it's well worth the effort.

Apply Loctite to the sway bar insulator mounting bolts when reinstalling them. Also use some anti-seize compound on the threads for the end links.

T1 Suspension Kit
The T1 suspension package runs about $3,000 for the total kit and $250 each for the shocks. It's a very straightforward spring, shock, control arm, and sway bar replacement. The kit includes front upper A-arms, front lower A-arms, front sway bar end links, and front and rear springs. Shocks must be purchased separately. You can get a kit from your GM dealer. If you go to www.gmpartsdirect.com it shows the pricing.

Front Bar Installation
(1) Use an 18-millimeter combination wrench and T40 Torx bit to remove the lower end link nut on both sides.
(2) Grab hold of one end of the sway bar and see if the bar

SUSPENSION

103

This plastic sway bar link was one of those ideas that didn't work out very well. It was fine as long as the car had the base suspension, but with the introduction of the Z06, a move was made to aluminum. Sway bar links are so easy to replace and so inexpensive, I would simply change them to aluminum on any early C5. Then put the plastic ones away in a safe place should you ever want to have the car judged at a National Corvette Restoration Society (NCRS) meet, or try to get a Bloomington Gold award.

easily rotates through its mounts on the frame. You should be able to do this with one hand and very little pressure. If you have to use significant force, the sway bar is binding in its mounts.

(3) If you have trouble rotating the bar while it is still installed in the brackets, check for rust buildup in the area between the bushings and the exterior of the sway bar.
(4) Remove the four bolts that hold the front sway bar to the frame, using a 15-millimeter socket.
(5) You can now place a new bushing on the sway bar. Make sure you install it the same way the stock one came off and lubricate the interior of the bushing with suspension lube.
(6) Apply a light coating of anti-seize compound to the four bolts that hold the bar to the frame.
(7) Using the original bracket, install the sway bar back onto the frame and start the bolts by hand. Don't tighten them at this time.

Rust forms between the rubber bushing and the steel sway bar on the C5. If your C5 is several years old, remove the bar from the car, push the bushing down the bar, and clean off the rust. Apply silicone lubricant to the area and put everything back into place. This will keep the bar rotating nicely in the chassis mounts.

> **Fixing Sway Bar Squeaks**
> When the temperature outside is 42 degrees Fahrenheit or colder, you might notice a squeak in the front and/or rear suspension. This squeak comes from the sway bar bushings. To fix the squeak, simply replace the bushings with new polyurethane bushings; they're cheap and it's an easy job.
> (1) Raise the car so that either the front (or rear) wheels are off the ground.
> (2) Remove the two bolts and the bracket that hold the sway bar bushing in place. Do this for both bushings. The rear sway bar bracket has one bolt and one nut.
> (3) Swing the sway bar down. Note the position of the bushing on the sway bar, and slide both bushings toward the center of the sway bar.
> (4) Clean the contact areas of the bushings on the sway bar. You may have to use sandpaper on an older car.
> (5) Apply a silicone lubricant to the inside of the bushings before moving the new bushings into place.
> (6) Place the bushings on the sway bar in the same position as the old bushings.
> (7) Move the sway bar back into position.
> (8) Position the bracket back in place so that the bushing fits inside the bracket and the tab on top of the bushing fits into the notch on top.
> (9) Start both bolts (or rear bolt and nut) back in to hold everything in place. Do this for both brackets.
> (10) Make sure the sway bar is centered.
> (11) Tighten each bolt down a little bit at a time so that the bracket fits on the bushing straight.

(8) Now you can connect the links.
(9) With that done, make sure the bar is centered and then tighten all the nuts and bolts.

Rear Bar Installation
(1) Use an 18-millimeter combination wrench and T40 Torx bit to remove the links.
(2) Try to rotate the bar through its range of motion to be sure nothing is binding.
(3) Once you've determined that everything is OK, you can use a 15-millimeter socket and an 18-millimeter wrench to remove the lower nut. Don't get carried away and remove the entire bolt, since that holds the lower A-frame in place.
(4) Push the lower ends of the links out and remove the bar from the car.
(5) Install the rear bushing the same way you installed the front, being careful to not confuse the factory original orientation. (Don't install it upside down.)
(6) Apply anti-seize compound to the two frame mount bolts, and install the bar back in its place.
(7) Make sure it's centered and then tighten everything to the appropriate specification.

PROJECT 36 | Selecting Tires and Wheels

Time: As long as it takes to make a decision

Tools: N/A   **Torque:** N/A

Talent: ★   **Applicable Years:** 1997–2004

Tab: You can spend up to $3,000 on tires and wheels.

Tip: Find a really good tire store to mount your tires and wheels. Or, find a wheel vendor that sells complete packages that are ready to simply bolt on your Corvette.

Performance Gain: Wider tires and softer rubber compounds will improve handling.

Complementary Work: Every time you have the wheels off the car, you should check for brake pad wear.

There are a huge number of wheels offered for the Corvette. Some are an improvement, but others are a step back in quality and strength. Start with the standards that stock GM wheels must meet and base your decision on that level of quality.

Strength

The strength of a wheel is first and foremost a safety consideration. If a wheel were to fail, it would likely happen at the most inopportune time and result in unmitigated catastrophe. Because of its overall performance envelope, a Corvette is capable of imparting much higher loads on its wheels than the average passenger car, and this must be factored into the equation. The strength of the Corvette's wheels also has a significant impact on the car's handling.

Before they are released for production, Corvette wheels undergo three primary evaluations conducted in accordance with Society of Automotive Engineering (SAE) specifications: the rotary fatigue test, radial fatigue test, and impact test. For the rotary fatigue test, wheels are mounted on a machine that simulates cornering under a very heavy load and at a very severe angle, both of which exceed what the wheel will ever see in actual service. The radial fatigue test also employs a load factor many times greater than what the wheel will encounter in actual use, but instead of cornering, the wheel is run straight ahead, and instead of being bare it has a tire mounted. Just as the name implies, the impact test drops very heavy weights onto the wheel while it's mounted in a special fixture to simulate the kinds of forces it's likely to experience in different types of crashes.

C5 chassis engineers worked very hard to create the optimum suspension and steering geometry that, together with proper wheel alignment settings, provides the most desirable

These are the base wheels from the 1997 C5. The most common upgrade for a few years was to simply have these wheels chromed. Now the wheel industry has gone way beyond that.

combination of handling, ride quality, noise abatement, and so on. Corvette wheels that deflect significantly under a heavy load will adversely affect handling and ride characteristics, adding even greater importance to selecting extremely strong wheels. In other words, cheap wheels are a really bad deal on the Corvette.

Mass

Getting weight out of a Corvette by reducing the mass of its wheels is a win-win-win situation. Reducing the car's overall weight helps to reduce fuel consumption, extend tire life, and

SUSPENSION

Tire Pressure Sensor Calibration Procedure
(1) Make sure that all four wheels are on the ground and properly tightened to specifications.
(2) Turn the ignition to the "on" position.
(3) Press the "DIC reset" button to clear any warning messages.
(4) Press the "DIC options" button until the display goes blank.
(5) Press and hold the "DIC reset" button for 3 seconds.
(6) Press the "DIC options" button again until the "tire training" message appears.
(7) Press the "DIC reset" button until the message "learn L front tire" appears.
(8) Hold the magnet over the pressure sensor's location on the left front wheel (this should be next to the valve stem if installed correctly).
(9) When the horn beeps, the programming is complete for that wheel.
(10) Program the other three wheel sensors as directed by the message display. The sequence is as follows: left front, right front, right rear, and left rear.
(11) For more information, refer to the Corvette service manual, page 3-114: Low Tire Pressure Sensing—TPM Sensor Programming.

improve acceleration, braking, and handling. Wheels are included in what's called unsprung weight. Reductions to unsprung weight pay extra dividends in the form of faster suspension movement, which allows the tires to maintain better contact with the road surface while enhancing overall stability. Wheels are also considered rotating inertia and, as with unsprung weight, reductions in this area pay extra dividends in the handling department.

The lightest wheels that were available on the C5 were the optional magnesium units that were first offered in 1998, discontinued in 2002, and offered again in 2003. Though the price for these dropped from $3,000 to $2,000 beginning in model year 2000, they're still relatively expensive, which accounts for their small sales numbers. A variation of this C5 magnesium wheel was also offered on the C6 Corvette in 2005.

The base and Z06 wheels are generally referred to as "cast" units, but they are actually only partially cast. The front face of each wheel is cast in a mold by a traditional casting process using A356 aluminum. The rim portion of the wheel is not cast, as it would have been in the past. Instead, beginning midway through model year 2000, the rim was made through a process called "flow forming." When the face is cast, a quantity of excess material remains around its circumference. This skirt of additional aluminum is heated and shaped into the rim by tooling that squeezes it into the desired configuration while it is being spun.

Corvette's optional polished-aluminum wheel, the face of which is made using a forging process, also relies on flow forming for its rim section. Tooling in very powerful presses strikes a T6061-T6 aluminum blank four times to forge the face, while leaving a skirt of extra material around the perimeter. As with cast wheels, flow forming is employed to massage the skirt into a rim. The process of flow forming, which is similar to working material on a lathe, yields better grain structure than is otherwise achievable and results in a stronger wheel. Achieving greater strength of material means less material has to be used to create otherwise equal wheels. Cast-spun and forge-spun wheels are clever innovations that embody the leading-edge technology found in today's Corvettes.

Domestic Standard Aluminum Five-Spoke—RPO QD4
Front: 17x8.5 inches, part number 9592413
Rear: 18x9.5 inches, part number 9592414

Export Standard Aluminum Five-Spoke—RPO QF3
Front: 17x8.5 inches, part number 9592613
Rear: 18x9.5 inches, part number 9592615

Export Standard Aluminum Five-Spoke—RPO QD4
Front: 17x8.5 inches, part number 9592413
Rear: 18x9.5 inches, part number 9592414

Export Standard Aluminum Five-Spoke—RPO QF3
Front: 17x8.5 inches, part number 9592613
Rear: 18x9.5 inches, part number 9592615

Domestic/Export Magnesium Five-Spoke—RPO N73**
Front: 17x8.5 inches, part number 9592638
Rear: 18x9.5 inches, part number 9592640
** These wheels always had the offset to meet export requirements.

The 1999 Corvettes came with standard five-spoke aluminum wheels; however, there were different part numbers for domestic versus export models. This means that there are actually six different part numbers for wheels for 1999. Magnesium wheels were optional for both domestic and export owners. Export wheels have a different offset at the mounting point, which brings the wheel and tire inboard under the fender well approximately ½ inch. You can identify these wheels by a black line drawn around the circumference of the wheel between the beads, seen only when the tire is removed from the wheel.

Many Corvette owners select wheels based on looks alone. Just use common sense and realize that in most cases the aftermarket wheels you're thinking of buying have not been subjected to the variety of tests that GM uses.

If you go to an open wheel design you're almost obligated to install large brake rotors and calipers.

SUSPENSION

PROJECT 37 | Performing an Alignment

Time: Several hours to a day if you do your own alignment

Tools: N/A   **Torque:** N/A

Talent: ★   **Applicable Years:** 1997–2004   **Tab:** $250

Tip: Paint yellow stripes on all of the bolts that should be moved in a normal alignment. That way you can see if the alignment shop actually changed anything on your car. I also do this before I take my car to the track since I can quickly see if a nut is loosening.

Performance Gain: Driving a car that's properly aligned is not only a real joy, but it adds to the total performance of the car, since it's not dragging its wheels down the street in the wrong direction.

Complementary Work: This might be a good time to install new shocks.

There are really two tests for your C5's alignment. First, does the car drive properly? Next, are your tires wearing evenly? These criteria apply whether you're driving over to visit your in-laws or running at Sebring. The settings may be different, but the criteria are the same.

We want the car to drive down the road in a straight line, and we want those big expensive tires to last as long as possible. A good alignment is essential for maximum traction and high-speed performance. You will only need to adjust the alignment when something causes your wheels to move from the last time you had an alignment. A giant pothole could cause shifts in your suspension. Also, wear in your suspension components can cause all the settings to change.

I recommend an annual preventive maintenance alignment job, but there are several other ways to decide if an adjustment is really necessary. First, check the tread depth of your tires on a regular basis. All four tires should wear evenly across the tread. Any Corvette, of any year, should go down a straight road without steering input. Steering wander at highway speeds might be a result of accumulated chassis wear, but it's a sure sign that your wheels are not pointed in the appropriate direction.

There are few people left in the service industry who can perform a proper alignment. Your Corvette needs an outstanding alignment specialist, whatever type of shop they work in. It isn't that the Corvette is so difficult to align; it's

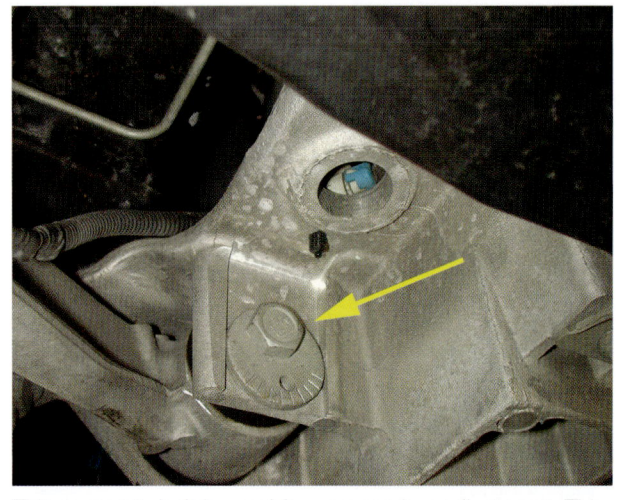

This eccentric bolt is used for rear camber adjustment. This part has been around since 1963, when the Corvette first had an independent rear suspension.

Front Alignment Specifications

Front Individual Toe	0.04 degrees
Front Total Toe	0.08 degrees
Front Individual Camber	-0.20 degrees
Front Cross Camber	Within 0.50 degrees
Front Individual Caster	6.9 degrees
Front Cross Caster	Within 0.50 degrees
Steering Wheel Angle	0.0 degrees

Rear Alignment Specifications

Rear Individual Toe	-0.01 degrees
Rear Total Toe	-0.02 degrees
Rear Thrust Angle	0.0 degrees
Rear Individual Camber	-0.18 degrees
Front Cross Camber	Within 0.5 degrees

Smart Racing Products makes one of the best camber gauges around. It's inexpensive and accurate. Since it's fairly easy to adjust the camber on your C5, why not play around with negative camber just prior to a track event or autocross? When you're done playing you can return to the stock camber setting.

This is called a toe plate. There is another plate on the other side of the car that leans against the wheel the same way. Measure the distance between the front edges of the two plates. Then measure the distance between the rear edges of the two plates. If the front edge measurement is greater than the rear measurement, then you have toe out. If the rear of the plate is a greater distance than the front, you have toe in. If you run Solo 2 events, you may want to play around with the toe settings. It only takes a few moments to change the toe, and it can be just as easily changed back.

SUSPENSION

109

Small protractors are available at home improvement stores. Remember that you want the nose of the C5 Corvette a little lower than the rear. I like to set it at 1 or 2 degrees.

just that most of us are very picky about the way we want our Corvettes to drive.

One way to identify a quality shop is to determine the type of equipment that's being used. You don't need the latest whiz-bang equipment, but you do need equipment that's well maintained. Alignment equipment is very fussy, and it needs to be treated properly.

The newest alignment equipment has to be extremely easy to use so the technician can do the job quickly and move on to the next customer. Most good technicians work on some sort of commission, and the time it takes to set up and align your Corvette has a direct impact on how much money they make each day. The latest equipment won't let them do a better job on your Corvette. It'll only let them do it faster. Ultimately, it's still the operator of the machine that makes the actual adjustments and decides what is good enough for your Corvette.

You may hear specialized vocabulary at the alignment shop. The following is intended to demystify those specialty terms.

Toe: The first word you're going to hear is "toe." This is a description of how straight the wheels go down the road. You want the right front tire going down the road parallel to the left front tire, and both of the tires should be parallel to the centerline of the car.

If the wheels are pointed toward the center of the car, we say that the wheel is "toe in." If the wheel is pointed to the outside of the car, we call it "toe out." If one of your front tires is pointing too far to the outside, or the inside, you'll wear that tire out quickly. The toe specification for the rear tires is just as important as for the front wheels.

Caster: "Caster" determines how well your Corvette goes down the road. You usually won't get poor tire wear from incorrect caster, but you will have steering pull. As a rule of thumb, the more caster you have in your Corvette, the easier it is to drive in a straight line. The tradeoff is increased steering effort, but with power steering and street driving, the extra caster shouldn't present a problem.

Camber: "Camber" is a term that tells us how the wheels relate to the pavement in a vertical plane. We want the wheels to be vertical as we go around a corner, so they may need to be slightly inclined as we drive down the road.

The C5 comes from the factory with a very slight amount of negative camber. That means the wheels tilt in at the top. This enhances the turning characteristics. Too much camber and you'll wear out the tires. Too little camber and not only will you wear out your tires, but the car won't corner properly.

When you change your alignment specifications from the factory recommendation, you should have a good reason for doing it. The same rule applies if the shop suggests a change.

If you change your tires and wheels, you're on your own for the alignment settings. The stock settings generally work nicely for a street-driven Corvette. If it's an autocross or track car, buy an accurate tire pyrometer and learn how to read tire temperatures.

PROJECT 38 | Replacing the Sway Bar End Links

Time: 1 hour

Tools: Sockets, screwdrivers, and drills

Torque: Rear stabilizer bar: bracket bolts: 61 N-m (49 ft-lb), bracket nuts: 95 N-m (70 ft-lb), end link nuts: 72 N-m (53 ft-lb); front stabilizer bar: insulator clamp bolts: 58 N-m (43 ft-lb), end link nuts: 72 N-m (53 ft-lb), end link nuts: 72 N-m (53 ft-lb)

Talent: ★★★★ **Applicable Years:** 1997–2004 **Tab:** $50 to $100

Tip: N/A

Performance Gain: Probably no gain in performance, but it's fun to do.

Complementary Work: You might want to install polyurethane bushings at this time.

The 2002 Z06 Corvette was equipped with aluminum rear sway bar end links. These aluminum links are easier to work with than the 1997–2001 end links. Most of the C5 aftermarket vendors stock the aluminum links, but you can also get them from a GM parts dealer. These end links also come with the quality aftermarket sway bar packages.

The GM 2002 aluminum end links retail for about $20.25 each, and you may need four of them. When ordering the 2002 end links, refer to GM part number 10435298. The same part number fits both sides. This is also the part number for the 2002 front sway bar end links. General Motors managed to use one part in four different locations—that's smart engineering.

Some of the early Z06 cars came with metal, but not aluminum, links from the factory. You may want to replace them with the aluminum version. If your Corvette front end links are plastic, you'll want to perform this modification to the front as well.

The 1997–2001 links use a Torx bolt (TX40) as a mounting bolt. In 2002, the part was changed to a metric Allen head key instead of the Torx drive. This new 2002 part also gives you the option of using an 18-millimeter open-end wrench on one end while using an 18-millimeter socket for the nut on the other end of the mounting bolt. They did this so that you can securely hold the bolt while tightening it down to the proper specification. This makes it much easier when mounting and tightening the 18-millimeter nut. You can mount them this way without even using an Allen key.

The first thing you need to do for this project is to raise the rear (or front) of the car. Place a board on the rear crossmember and get the entire rear of the car off the ground. Use two jack stands for support. You don't have to remove the rear wheels, but it makes the job easier.

These plastic sway bar end links are of use only to the restoration folk. Replace the end links with the aluminum version but remember to keep the plastic ones around should you ever want to restore the car to its original condition.

Start by removing the 18-millimeter nut. Insert the Torx drive into the end of the end link bolt to hold it securely while you break the nut loose with an 18-millimeter open-end wrench. While holding the bolt stationary with the Torx drive, continue to loosen the nut with the 18-millimeter open-end wrench until you can remove it with your fingers. Don't remove the nut until you've repeated this same procedure with the lower nut and bolt.

For some reason, the upper nut is harder to loosen and remove than the lower, although the lower one is a little tougher to get to. Once you get the nuts loose enough, remove both of them. Keep these nuts handy, as you will be reusing all of them. Now, simply remove the plastic end link.

SUSPENSION

111

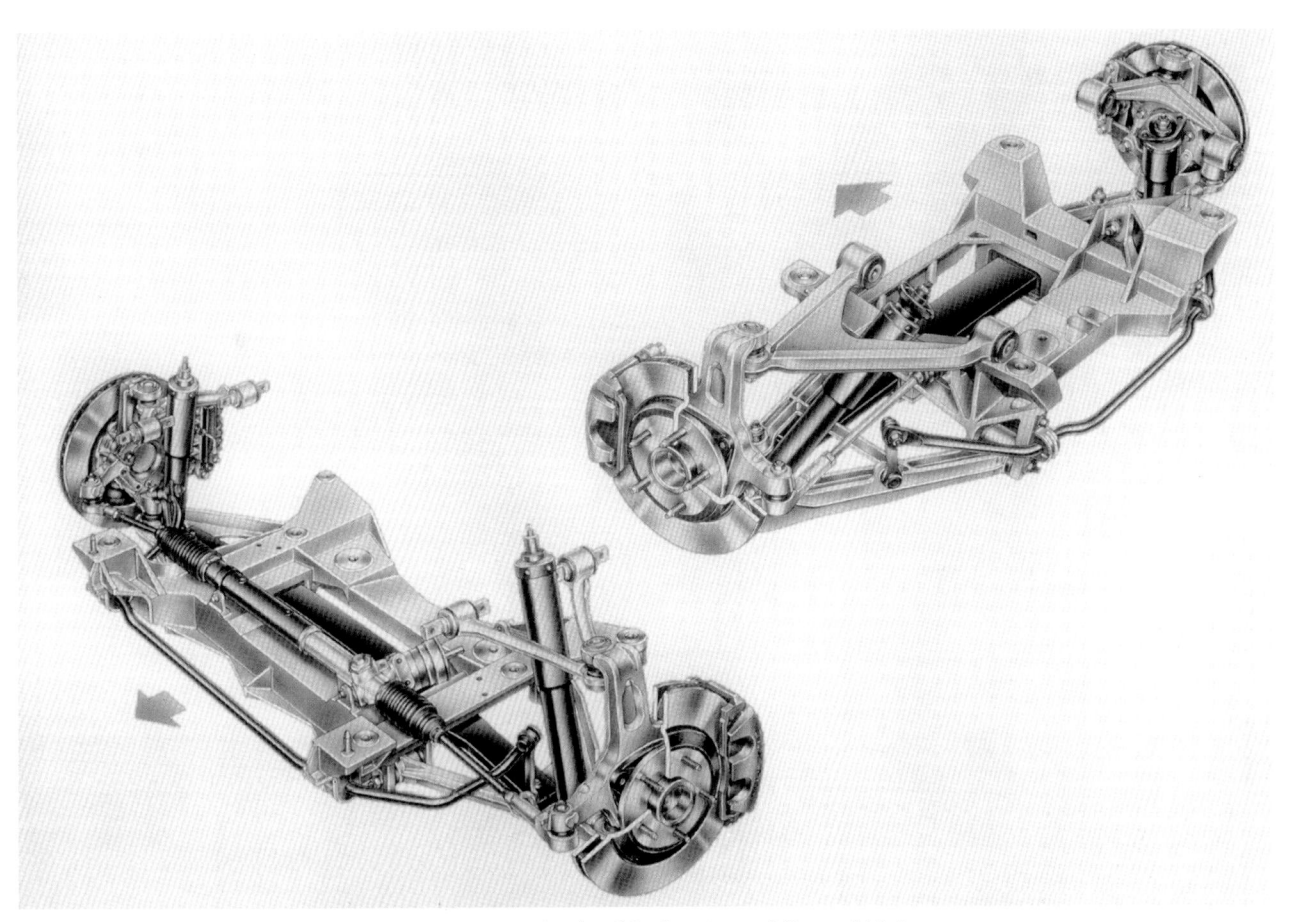

These drawings illustrate the sway bar arrangement for the C5. *Courtesy of General Motors*

You may have to work with it for a few seconds, but you should be able to remove the end link fairly easily.

You'll need to make a decision: If you're installing polyurethane insulator bushings, remove both end links and then remove the supports that hold the sway bar to the rear crossmember. Drop the entire sway bar down off the car and put it on your garage floor.

Whether using new polyurethane bushings or simply putting the original ones back in place, make sure you lube the inside of the bushings. If you're going to stay with the stock bushings, slide them down the bar and grease the area where the factory bushings will be located when you're done. There should be no friction between the bushings and the sway bar.

When you bolt the sway bar back onto the crossmember, and before you attach the end links, move the sway bar through its range of motion. It should move easily. If it takes any real effort to move the sway bar, sand the area of the bar that is in contact with the bushing. Once it moves easily, you're ready for the new end links.

Apply some anti-seize compound to the threads of the new aluminum end link bolts. Install the new aluminum end link bottom (lower control arm) bolt first. Put the 18-millimeter nut on the bolt until finger tight. Swing the end link around and up to put the upper portion through the sway bar end. You may have to move the sway bar a little bit, but it will move fairly easily without much effort. Put on the 18-millimeter nut until finger tight.

You'll notice that the aluminum end link is a much nicer piece and is very easy to install. Use your 18-millimeter open-end wrench on the base of the end link bolt where GM designed a spot to grip the bolt while tightening/loosening. Hold the bolt with the 18-millimeter open-end wrench. Use a ½-inch drive 18-millimeter deep socket and ratchet to tighten the nut. Tighten the bolt down to 53 ft-lb of torque.

You'll probably find that once you've completed one side, the other side will take half as long to do. Repeat all of the above steps for the opposite side. When you're finished, double-check everything. Carefully lower the car and take it for a drive. There should be no squeaks or rattles coming from under the rear (or front) of the car. If you hear anything, chances are something is wrong and you need to recheck your installation to find the problem(s).

From start to finish, this entire procedure shouldn't take longer than 1 ½ to 2 hours to complete, but don't get in a hurry. Take your time and do it right.

SECTION SEVEN
COSMETICS

Projects 39–40

PROJECT 39	Customizing the Interior Trim

Time: A few hours **Tools:** N/A **Torque:** N/A

Talent: ★★ **Applicable Years:** 1997–2004 **Tab:** $50 to $500

Tip: Always check out how some of these items actually look in a real car before you order your parts. The quality can vary widely.

Performance Gain: N/A

Complementary Work: If you're going to this much trouble to make the interior special, make sure you detail the outside of your Corvette as well.

There's a tremendous amount of stuff you can put in the interior of your C5 to personalize your Corvette. Some people want a race car look and others want a luxury cruiser. It's hard to believe, but a lot of people prefer listening to a sound system to a set of straight pipes.

Here are installation techniques for some of the most popular modifications:

Dead Pedal Installation

Remove the floor mat from the driver's side and locate the 10-millimeter nuts that hold the factory dead pedal in place. This is most easily done if you take out the bottom bolt first.

Place the new aluminum dead pedal on top of the factory assembly and mark the position of the holes on the plastic factory piece. Drill two 6-millimeter holes. These holes will be a little too small for the machine screws that came with the aluminum pedal. Use a file to enlarge the holes as necessary. The ribbed surface of the factory piece makes it difficult to drill the holes exactly where you need them, so take your time.

Align the aluminum dead pedal on the factory dead pedal and adjust the bottom hole with the file until it lines up. Then insert one of the supplied machine screws to hold it in place, and adjust the top hole with the file.

Using the 11-millimeter machine screws and nuts, screw the aluminum dead pedal onto the factory unit. You can use a screwdriver to hold the screws in place and tighten the nuts using an 11-millimeter deep socket.

The interior panels of the C5 are easy to remove, which has allowed the aftermarket to create some really nice replacement panels.

Now replace this assembly unit in the car and torque the nuts to 10 N-m. You'll need an extension and socket on the top one, which can be difficult to install.

Gas Pedal Installation

Remove the factory gas pedal by first lifting the spring off the connecting metal lever to remove the tension. Spread the spring while pulling at the pin that holds the factory gas pedal. The pin is held in place by the spring, which sits in a groove in the center of the pin.

Bend (open) the factory spring enough to allow the screw on the new aluminum gas pedal to pass through the center

The A-pillar has become a very popular place for additional gauges. The best part is that they are easy to remove when the time comes to sell your car. A lot of parts like this will actually lower the resale value of a Corvette.

The center console is a wonderful place to personalize your Corvette.

hole. This makes installation much easier, as the new screw does not have a groove in the center.

Once you are sure that the mounting screw can pass easily through the center of the spring, put the aluminum gas pedal on the metal lever and push the screw through from left to right. The spring needs to go on the right side of the lever, between the lever and the right edge of the aluminum gas pedal with the screw going through the center. Make sure the spring is oriented correctly and fasten the screw using a 2-millimeter Allen wrench.

Bend the spring back over the metal lever, creating tension on the pedal, and step back and enjoy your new pedal.

Shift Knobs
This is one of those Saturday-morning jobs that I love so much. You'll spend only about an hour out in the garage and less than $50.

Since you're going to have the car in neutral while you do this job, block the tires to keep the car from moving.

Move the shifter into neutral or drive.

There is a C-type clip in the front of the stock shift knob almost to the bottom that you will have to pry out to lift off the stock knob. You can see it sticking out after you've started to remove it. Use a very small screwdriver to remove the clip.

Now you'll have to take the leather boot out from under the plastic washer. You can pry the leather boot off with needle-nose pliers, working the leather from under the washer.

There's a plastic clip holding the leather in, and you'll have to pull out the leather by working around the shaft.

Once you've worked the leather and its plastic clip out, push the leather boot down the shaft.

Place the spring on the shaft before you put on the adapter. The spring will help lift the shift ball between shifts.

Place the adapter on the shaft, and move it around until it slides down. (It will initially sit on top of the spring.)

Tighten the two Allen setscrews with an Allen wrench. The Allen screws tighten on the rod that comes up from the shaft.

Push the leather boot up over the larger part of the adapter.

Next, place the collar over the top of the leather boot. You'll have to work the leather under and the collar over. It helps to turn the collar as you are working it. Push the collar down as far as possible.

Depending on the aftermarket knob or ball you use, it may be higher than your stock knob by about ½ inch. You can adjust this a little by drilling most knobs or balls out deeper.

Some knobs, such as the MOMO, have a sleeve that fits into the knob and are then held on by Allen setscrews. You can also try wrapping the shaft with electrical tape until the appropriate thickness is achieved.

Insulation Installation
Corvette C5 insulation kits are designed to reduce the drivetrain heat and road noise that radiate through the floor and into the cabin. If you have a convertible or a Z06, be warned that most of the kits, such as KoolMat, are based on the coupe and require a fair amount of trimming and cutting to get a proper fit. Even so, the precut kit is easier than trying to cut it yourself from a roll of material.

The best way to install the kit is to just start gutting the entire interior. This means seats, carpeting, and plastic side moldings.

It even helps to unbolt the trunk lid latch so you won't have to cut around it.

Once the rear interior is empty and clean, test-fit the pieces in the specified pattern. Most of the precut kit pieces are close, but they never seem to be an exact fit.

Once you've gotten everything out of the car, protect everything that's left against overspray. Then spray the contact adhesive on both the interior panels and the fiberglass side of

Most of the early C5 criticism focused on the interior because it was obvious that GM was cutting corners. The same company manufactured the seats for both the Corvette C5 and the BMW M3, but BMW paid twice as much for the seats, and it showed. When the C6 was introduced, GM spent a tremendous amount of time telling everyone how much effort had gone into improving the interior.

the thermal barrier. Wait a few minutes for each section to become tacky.

Work your way across, one piece at a time, while trimming as necessary to fit them smoothly together.

Once the floor sections are laid down, you can trim and glue the wheelwell panels into place.

Take your time and trim each piece until it fits snugly.

You can carve and shape pieces of the kit to fit into the side cargo compartments. Most kits include plenty of material to work with.

After locating and marking bolt locations on the rear wall panel, use a punch to cut out precise holes to fit.

The back wall panel usually comes as a single piece, but there are enough complicated curves and obstacles that you might decide to cut it into several sections for the best overall fit.

You're also going to have to trim the piece for the passenger's side of the rear bulkhead to fit cleanly around the electric antenna wiring harness.

On the driver's side, you'll find a hole for the body pressure relief valve. If you accidentally seal this opening off, it's going to become extremely difficult to close the doors.

The center floor storage area is a nightmare of complex angles. Simply work with small pieces to fill in the little voids with properly trimmed scraps of material.

Apply seam sealer to all of the seams. Once it is fully cured, you can reinstall the carpeting, including the factory carpet pad.

Remove the front seats, the kick panels, and the lower seatbelt brackets. Then pull out the carpet.

Remove the two-piece console next.

With the console out, remove the carpeting that covers the sides of the torque tube tunnel.

Once the front compartment is empty, lay the floorboard insulation in place, mark the locations of the seat bolts on the insulation panels, and punch out the necessary holes at your workbench.

You're going to have to cut slits in the insulation along the rocker panel for the wiring harness clamps to slip through.

Slit the tunnel shifter piece so that it can slip over the shift knob. It's easiest if you remove the shift boot to do this.

For the cleanest installation possible, unbolt the emergency brake lever from the tunnel. Only remove the three bolts that secure it, leaving the cable intact so you don't have to readjust the brake cable later.

Now turn your attention to the shifter. You don't want to lose future access to the shifter, so unbolt the shifter cover plate and cut the insulation so that it fits right up to the flange on the tunnel underneath it.

Bolt the shifter cover plate back down on top of the insulation.

With everything in place, you can start putting in the carpet. This is a great time to install new carpeting.

COSMETICS

Gauge faces can be easily changed in the C5.

This is the most functional C5 dash in the world. It's the center dash panel on the C5-R.

For less than $200, these trim rings truly give you a unique look.

COSMETICS

PROJECT 40 | Installing a Body Package

Time: 6 hours

Tools: Sockets, screwdrivers, and drills

Torque: N/A

Talent: ★★★★

Applicable Years: 1997–2004

Tab: $500 to $10,000

Tip: Talk to a Corvette owner who has actually installed the spoiler you're thinking about putting on your Corvette.

Performance Gain: Probably no gain in performance; in fact, many spoilers may actually detract from top speed.

Complementary Work: You may want to replace the rocker panel when you add an aftermarket spoiler.

The price you see in the catalog is only for the spoiler—very raw and usually very unfinished. A basic wing is going to cost from $200 to $500. That means you have a wing that arrives at your home in unfinished condition and fits reasonably well. You'll have to invest some money to get the wing finished and installed on your car. Simply finding a body shop that is skilled in working with fiberglass to do the job can be challenging.

Not only will these body shops charge you a substantial amount of money, but many will even complain about having to perform the work, because insurance work is their bread and butter.

The best way to select a spoiler is to attend car shows and find one that looks right to you. None of them, except for the very rare factory spoiler, actually do any real good. The spoilers that you see in all the catalogs are simply cosmetic items. The only spoiler that really adds downforce to the rear of your car is the original factory spoiler. But the spoiler was unpopular and Chevrolet quickly discontinued it from production, and most of the fiberglass companies no longer manufacture it.

Over the years, the market has not treated Corvette body modifications kindly. Adding a spoiler will likely decrease the value of your car relative to the cost of the modification. If you spend $1,000 on the spoiler, you'll take at least $500 off

This is one of the more tasteful spoilers for the C5. It probably adds very little downforce, but it looks nice. None of the aftermarket spoilers have ever been inside a wind tunnel. It's all about the look.

These aftermarket rear brake vents look so right on the Corvette. But they require cutting holes in the body, and not many people seem to want to do that.

COSMETICS

Screens give the C5 the aggressive look that it lacked in 1997. And since they don't require destruction of any body panels, they won't detract from the value of your car.

the value of your car. Most spoilers require that you drill holes in the body of the car, and that makes it less attractive to prospective buyers down the road.

The C5 body panels, however, are bolted on. This means you can install wider rear fenders and then put the stock ones back on when it's time to sell the car.

Screening

Putting a window screen in every possible opening of your C5 Corvette has almost become a cliché. The screening looks good, doesn't cost a lot of money, and is easy to install. You can make your own screening or buy it through an aftermarket catalog. Just be sure your screening is made from stainless steel.

The aftermarket screens are pressure-fitted between the plastic bumper and the ridged brake duct, or other openings. The best-looking screens are installed from behind the bumpers.

If you're going to push the screens in from the front, you should cover the openings completely with duct tape. Trying to install the screens without the tape will result in scratch marks since the screens are bigger than the openings. Once the tape is applied, just use a little elbow grease and bend the screens to install them.

The best installation method, however, involves going in from behind and removing some plastic panels. It takes a little longer, but this method will prevent scratching. The screens are pressure-fitted using a plastic accordion-style brake duct to hold them in place.

Hoods

Hoods are easy to install, easy to paint, and easy to return to stock, but they aren't cheap. And the price of the hood itself may only be half of the money you'll spend on the job.

You have a wide variety of hoods to choose from, and when it's time to sell the car you can always put the stock hood back in place. Remember, in 25 years, the stock Corvette will always be the most valuable one.

These aftermarket hoods all take a considerable amount of work to look decent, and their quality is far below what's already on your Corvette. The average body shop hates working on them, so you end up paying a premium price.

Body Side Moldings

Chuck Mallett (Mr. Inside GM Powertrain) did a few wind tunnel tests with Chevrolet at GM's wind tunnel facilities in Troy, Michigan, when GM was developing the World Challenge body package and the C5 Aero kit. They noticed

Something as simple as these chrome rings can give you a personalized look.

a difference in rear end lift (or downforce, depending on your perspective) between cars with and without the body side moldings. They found that cars with the body side molding, or spears, had better rear downforce numbers and cars without side spears had less rear end lift. The spears aid and improve laminar flow along the sides of the car, and that improved laminar flow prevents the air from spilling off the side of the hood at the cowl. This in turn adds downforce at the windshield.

The improved side body laminar flow also reduces the amount of air wrapping under the car (in front of the rear wheels) and thus reduces rear end lift. Without the side spears, the reduced laminar flow and increased hood cowl spill almost doubles the amount of air getting under the rear.

Laminar flow only starts to become effective around 100 miles per hour and increases after that. There's a small downforce component of the side moldings due to their angle of attack. Although not measured, it can be assumed that laminar flow increases air extraction of the front and rear wheelwells, improves brake cooling, improves air extraction from functional side vents, and improves underhood air extraction from the evacuator hoods. This may not matter enough to justify putting the side spears on your C5. On the other hand, if they're already there, you now have a reason to leave them in place. This concept of improved side laminar flow was designed into the C6 and C6 Z06, and they therefore don't require the side moldings.

As time goes on, you may notice a large number of chips and dings on the lower side of the spears. Had they not been there, the stones and road grit would have been thrown up on the upper door, upper fender, top of the rear fascia, and the trunk lid, and could have promoted scratches, chips, and dings on these more conspicuous places.

Someone took and inverted the Lexan mounting strip for a Speed Lingerie bra (when not using it on the track) so that it stuck out of the wheelwell. They discovered that brake temperatures were reduced 100–200 degrees, front downforce increased, side laminar flow was not affected, and drag was not significantly increased. And although it was never measured, it can be assumed that the negative pressure generated behind the front fender lip increases cooling air to the brake through the ducts, cooling air entering the well from the bottom, ground effect in the front, air extraction from the wheelwell, and air extraction from a functional side vent.

The C5-R hood hasn't been copied much. Could this be the next big thing in C5 body kits? I just hope whoever does it includes the vents on the top of the fenders.

COSMETICS

Appendix

VIN Number Decoding

The Vehicle Identification Number (VIN) decodes as follows for the C5 Corvette. The last six positions show the car's position in the build sequence for the model year.

1G1YY31G935100001
1—Country of origin (1=United States)
G—Manufacturer (G=General Motors)
1—Division (1=Chevrolet)
YY—Carline and series (YY=Corvette)
3—Body type (1=hardtop, 2=two-door coupe, 3=two-door convertible)
1—Seatbelt system (2=active, manual w/driver and passenger inflatable restraints)
G—Engine code (G=LS1, S=LS6)
9—Check digit
3—Model year (see chart below)
5—Factory code (5=Bowling Green, Kentucky)
100001—Plant sequential build number

Model Year Chart
(This will be the 10th digit in the VIN number.)

Designator	Model Year
V	1997
W	1998
X	1999
1	2001
2	2002
3	2003
4	2004
5	2005

Regular Production Options (RPO)

1997

Code	Description
AAB	Memory package
AG2	Power passenger seat
AQ9	Sport seats
B34	Floor mats
B84	Body side moldings
CC3	Removable roof panel, blue tint
CF7	Base roof
C2L	Dual removable roof panels
CJ2	Dual-zone air conditioning
C60	Manual A/C
CV3	Mexico
D42	Security shade
EXP	Export
FE1	Base suspension
FE3	Sport suspension (included with Z51)
F45	Selective real-time damping, electronic
G92	Performance axle ratio, 3.15:1
MN6	Six-speed manual transmission
MX0	M30 automatic transmission
NG1	Massachusetts/New York emissions
R8C	Customer pickup
T96	Fog lamps
UN0	Stereo system Delco-Bose with CD
U1S	Remote compact disk changer
V49	Front license plate frame
YF5	California emissions
Z49	Canadian options
Z51	Performance handling package

1998

Code	Description
Z4Z*	Indy Pace Car with MXO
Z4Z*	Indy Pace Car with six-speed
1SB	Preferred equipment group
AAB	Memory package (Req. CJ2)
AG2	Power passenger seat
AQ9	Sport seats
B34	Floor mats
B84	Body side moldings
CC3	Removable roof panel, blue tint
CF7	Base roof
C2L	Dual removable roof panels
CJ2	Dual-zone air conditioning
C60	Manual A/C
CV3	Mexico
D42	Luggage shade and parcel net
EXP	Export
FE1	Base suspension
FE3	Sport suspension (included with Z51)
FE9	Federal emissions
F45	Selective real-time damping, electronic
G92	Performance axle ratio, 3.15:1, MXO Auto
GU2	Standard axle ratio, 2.73:1, MXO Auto
GU6	Standard axle ratio, 3.42:1, six-speed
JL4	Active suspension (after 12/15/97)
LS1	Standard 5.7-liter SFI aluminum V-8
MN6	Six-speed manual transmission
MX0	M30 automatic transmission
NG1	Massachusetts/New York emissions
N73	Sport magnesium wheels
PA6	Indy wheels
QD4	Standard factory aluminum wheels
R8C	Customer pickup at the NCM
T96	Fog lamps
UN0	Stereo system Delco-Bose with CD
U1S	Remote compact disk changer
V49	Front license plate frame
XGG	Front tire P245/45ZR17 BW SBR

Code	Description
YGH	Rear tire P275/40ZR18 BW SBR
YF5	California emissions
Z49	Canadian options
Z51	Performance handling package (included with FE3 sports suspension)

1999

Code	Description
1SB	Preferred equipment group
AAB	Memory package (Req. CJ2)
AG2	Power passenger seat
AQ9	Sport seats
B34	Floor mats
B84	Body side moldings
CC3	Removable roof panel, blue tint
CF7	Base roof
C2L	Dual removable roof panels
CJ2	Dual-zone air conditioning
C60	Manual A/C
CV3	Mexico
D42	Luggage shade and parcel net
EXP	Export
FE1	Base suspension
FE3	Sport suspension (included with Z51)
FE9	Federal emissions
F45	Selective real-time damping, electronic
G92	Performance axle ratio, 3.15:1, MXO Auto
GU2	Standard axle ratio, 2.73:1, MXO Auto
GU6	Standard axle ratio, 3.42:1, six-speed
JL4	Active suspension
LS1	Standard 5.7-liter SFI aluminum V-8
MN6	Six-speed manual transmission
MX0	M30 automatic transmission
NG1	Massachusetts/New York emissions
N37	Power tilt, telescopic steering wheel
N73	Sport magnesium wheels (late production)
QD4	17-inch standard wheels, fixed-roof coupe
R8C	Customer pickup at the NCM
T82	Twilight sentinel
T96	Fog lamps (not available on fixed-roof coupe)
UN0	Stereo system Delco-Bose with CD
U1S	Remote compact disk changer (all models)
UQ5	Extended-range Bose speakers, fixed roof optional
UV6	Heads-up display
UZ6	Standard CD four-speaker, mast antenna, fixed-roof coupe
V49	Front license plate frame
XGG	Front tire P245/45ZR17 BW SBR
XYF	Tire fixed-roof coupe, front 225/50-R17
YYR	Tire fixed-roof coupe, rear 245/50-R17
YGH	Rear tire P275/40ZR18 BW SBR
YF5	California emissions
Z19	Gymkhana/performance package, fixed-roof coupe
Z49	Canadian options
Z51	Performance handling package (included with FE3 sports suspension)

2000

Code	Description
AAB	Memory package (Req. CJ2)
AG1	Power six-way driver's seat (hardtop only)
AG2	Power passenger seat
AP9	Parcel net (hardtop only)
AQ9	Sport seats (Req. AG2)
AR9	Bucket, leather-trimmed seating surface
B34	Floor mats
B84	Body side moldings
CC3	Removable roof panel, blue tint
CF7	Base roof
C2L	Dual removable roof panels
CJ2	Electric dual-zone air conditioning
C60	Manual A/C
CV3	Mexico export
DD8	Light-sensitive inside rearview mirror
DD0	Light-sensitive OSRV mirrors
D42	Luggage shade and parcel net
EXP	Export option
FE1	Base suspension
FE3	Sport suspension (included with Z51)
FE9	Federal emissions
F45	Selective real-time damping, electronic
G92	Performance axle ratio, 3.15:1, MXO Auto
GU2	Standard axle ratio, 2.73:1, MXO Auto
GU6	Standard axle ratio, 3.42:1, six-speed
JL4	Active handling
LS1	Standard 5.7-liter SFI aluminum V-8
MN6	Six-speed manual transmission (w/1YY37 $0)
MX0	M30 automatic transmission
NB8	California/Northeast emissions override
NC7	Federal emissions override
NG1	Massachusetts/New York emissions requirements
N37	Steering column, power telescope manual tilt
N73	Sport magnesium wheels
R8C	Customer pickup at the NCM
QD4	Domestic standard five-spoke wheel
QF5	Deluxe high-polish wheel
T82	Twilight sentinel
T96	Fog lamps
TR9	Lighting: cargo, underhood, vanity (hardtop only)
UN0	Stereo system Delco-Bose with CD
U1S	Remote compact disk changer (all models)
UV6	Heads-up display
UZ6	Bose speaker and amplifier system (hardtop only)
V49	Front license plate frame
XGG	Front tire P245/45ZR17 BW SBR
YGH	Rear tire P275/40ZR18 BW SBR
YF5	California emissions
Z19	Gymkhana/autocross package hardtop (includes MN6 and Z51 combined with XGG and YGH wheels, GU6 axle)
Z49	Canadian options
Z51	Performance handling package, larger-diameter sway bars than two previous years (included with FE3 sports suspension, includes power steering fluid cooler)

2001

Code	Description
AAB	Memory package
B34	Floor mats
B84	Body side moldings
C2L	Dual removable roof panels (coupe only)
CC3	Removable roof panel, blue tint
DD0	Electronic monochromatic mirrors
F45	Selective real-time damping (not available on Z06)
FE1	Base suspension
FE3	Sport suspension (included with Z51)
FE9	Federal emissions
G92	Performance axle ratio, 3.15:1, automatic transmission
GU2	Standard axle ratio, 2.73:1, MXO automatic transmission
GU6	Standard axle ratio, 3.42:1, six-speed transmission
LS1	Standard 5.7-liter SFI aluminum V-8
MM6	Six-speed manual transmission (M12 standard on Z06)
MX0	M30 automatic transmission
N37	Steering column, power telescope w/manual tilt
N73	Sport magnesium wheels (not available on Z06)
NB8	California/Northeast emissions override
NC7	Federal emissions override
NG1	Massachusetts/New York emissions
R6M	New Jersey surcharge (mandatory in New Jersey)
R8C	Delivery at National Corvette Museum (NCM)
QD4	Domestic standard five-spoke wheel
QF5	Deluxe high-polish aluminum wheel (not available on Z06)
U1S	Remote compact disk changer
UN0	Stereo system Delco-Bose with CD (standard on Z06)
V49	Front license plate frame
XGG	Front tire P245/45ZR17 BW SBR
YF5	California emissions
YGH	Rear tire P275/40ZR18 BW SBR
Z49	Canadian options
Z51	Performance handling package (not available with F45 or Z06)

2002

Code	Description
AAB	Memory package (Req. CJ2)
AAB	Memory package (Req. CJ2)
AG1	Power six-way driver's seat (standard on Z06)
AN4	Child seat tether
AP9	Parcel net (coupe only)
B34	Floor mats
B84	Body side moldings
B84	Body side moldings
CC3	Removable roof panel, blue tint
C2L	Dual removable roof panels
CV3	Mexico export
DD0	Electronic monochromatic mirrors
EXP	Export option
FE1	Base suspension
FE3	Sport suspension (included with Z51)
FE9	Federal emissions
F45	Selective real-time damping, electronic (not available with Z06)
G92	Performance axle ratio, 3.15:1, MXO automatic (not available with Z06)
GU2	Standard axle ratio, 2.73:1, MXO automatic
GU6	Standard axle ratio, 3.42:1, six-speed
LS1	Standard 5.7-liter SFI aluminum V-8
MN6	Six-speed manual transmission (with 1YY37 $0)
MN6	Six-speed manual transmission (with 1YY37 $0)
MX0	M30 automatic transmission (includes G92 performance axle ratio)
NB8	California/Northeast emissions override
NC7	Federal emissions override
NG1	Massachusetts/New York emissions requirements
N37	Steering column, power telescope manual tilt
R8C	Customer pickup at the NCM
R6M	New Jersey surcharge (mandatory in New Jersey)
QD4	Domestic standard five-spoke wheel
QF5	Deluxe high-polish wheel (not available with Z06)
QF5	Deluxe high-polish wheel (not available with Z06)
UN0	Stereo system Delco-Bose with CD
U1S	Remote compact disk changer (all models)
V49	Front license plate frame
XGG	Front tire P245/45ZR17 BW SBR
YGH	Rear tire P275/40ZR18 BW SBR
YF5	California emissions
Z49	Canadian options
Z51	Performance handling package (not available with F45 or on Z06; included with FE3 sports suspension, includes power steering fluid cooler)

2003

Code	Description
AAB	Memory package (Req. CJ2)
AG1	Power six-way driver's seat (standard)
AN4	Child seat tether
AP9	Parcel net (coupe only)
B34	Floor mats (standard)
B84	Body side moldings
CC3	Removable roof panel, blue tint
C2L	Dual removable roof panels
CV3	Mexico export
DD0	Electronic monochromatic mirrors
EXP	Export option
FE1	Base suspension
FE3	Sport suspension (included with Z51)
FE9	Federal emissions
F55	Magnetic selective ride control (not available with Z06)
G92	Performance axle ratio, 3.15:1, MXO automatic (not available with Z06)
GU2	Standard axle ratio, 2.73:1, MXO automatic
GU6	Standard axle ratio, 3.42:1, six-speed
LS1	Standard 5.7-liter SFI aluminum V-8
MN6	Six-speed manual transmission (with 1YY37 $0)
MX0	M30 automatic transmission (includes G92 performance axle ratio)
NB8	California/Northeast emissions override
NC7	Federal emissions override
NG1	Massachusetts/New York emissions requirements
N37	Steering column, power telescope manual tilt (not available on Z06)
N73	Magnesium wheels

Code	Description
R8C	Customer pickup at the NCM
R6M	New Jersey surcharge (mandatory in New Jersey)
QD4	Domestic standard five-spoke wheel
QF5	Deluxe high-polish forged wheel (not available with Z06)
UL10	AM/FM cassette, Bose
UN0	Stereo system Delco-Bose with CD (standard)
U1S	Remote compact 12-disk changer (all models)
V49	Front license plate frame
XGG	Front tire P245/45ZR17 BW SBR
YGH	Rear tire P275/40ZR18 BW SBR
YF5	California emissions
Z49	Canadian options
Z51	Performance handling package, larger-diameter sway bars than previous years (included with FE3 sports suspension, includes power steering fluid cooler; not available with F55 or on Z06)

2004

Code	Description
AAB	Memory package (Req. CJ2)
AG1	Power six-way driver's seat (standard)
AN4	Child seat tether
AP9	Parcel net (coupe only)
B34	Floor mats (standard)
B84	Body side moldings
CC3	Removable roof panel, blue tint
C2L	Dual removable roof panels
CV3	Mexico export
DD0	Electronic monochromatic mirrors
EXP	Export option
FE1	Base suspension
FE3	Sport suspension (included with Z51)
FE9	Federal emissions
F55	Magnetic selective ride control (not available with Z06)
G92	Performance axle ratio, 3.15:1, MXO automatic (not available with Z06)
GU2	Standard axle ratio, 2.73:1, MXO automatic
GU6	Standard axle ratio, 3.42:1, six-speed
LS1	Standard 5.7-liter SFI aluminum V-8
MN6	Six-speed manual transmission (with 1YY37 $0)
MX0	M30 automatic transmission (includes G92 performance axle ratio)
NB8	California/Northeast emissions override
NC7	Federal emissions override
NG1	Massachusetts/New York emissions requirements
N37	Steering column, power telescope manual tilt (not available on Z06))
N73	Magnesium wheels
R8C	Customer pickup at the NCM
R6M	New Jersey surcharge (mandatory in New Jersey)
QD4	Domestic standard five-spoke wheel
QF5	Deluxe high-polish forged wheel (not available with Z06)
UL10	AM/FM cassette, Bose
UN0	Stereo system Delco-Bose with CD (standard)
U1S	Remote compact 12-disk changer (all models)
V49	Front license plate frame
XGG	Front tire P245/45ZR17 BW SBR
YGH	Rear tire P275/40ZR18 BW SBR
YF5	California emissions
Z49	Canadian options
Z51	Performance handling package (included with FE3 sports suspension, includes power steering fluid cooler; not available with F55 or on Z06)

Product Recalls

1997
NHTSA campaign ID number: 97V044000
Recall date: March 20, 1997
Component: suspension: rear
Potential number of units affected: 1,414
Summary: The rear suspension tie rod assembly can fracture.
Consequence: If this were to occur while the vehicle was in motion, loss of directional control can result, increasing the risk of a vehicle crash.
Remedy: Dealers will replace the left- and right-hand rear suspension tie rod links.
Notes: Owner notification began March 31, 1997. GM issued a "stop sale" to its dealers on March 20, 1997. Owners who take their vehicles to an authorized dealer on an agreed-upon service date and do not receive the free remedy within a reasonable time should contact Chevrolet at 1-800-222-1020. Also contact the National Highway Traffic Safety Administration's auto safety hotline at 1-800-424-9393.

1998
NHTSA campaign ID number: 00V111000
Recall date: April 6, 2000
Component: seatbelts: front
Potential number of units affected: 71,569
Summary: Vehicle description: passenger vehicles. The lap-belt webbing can twist, allowing the webbing to become jammed in the retractor.
Consequence: When the belt webbing becomes jammed in the retractor, the seatbelt may be unusable.
Remedy: Dealers will install inserts to the belt web guide of each lap-belt retractor.
Notes: Owner notification began November 13, 2000. Owners who take their vehicles to an authorized dealer on an agreed-upon service date and do not receive the free remedy within a reasonable time should contact Chevrolet at 1-800-222-1020. Also contact the National Highway Traffic Safety Administration's auto safety hotline at 1-888-327-4236.

NHTSA campaign ID number: 04V060000
Recall date: February 6, 2004
Component: Steering: column locking: antitheft device
Potential number of units affected: 126,624
Summary: On certain passenger vehicles equipped with electronic column-lock systems (ECL), when the ignition switch is turned to "lock," the ECL system prevents turning of the steering system. When the vehicle is started, the ECL unlocks the steering system.

The vehicle is designed so that if the column fails to unlock when the vehicle is started and the customer tries to drive, the fuel supply will be shut off so that the vehicle cannot move when the vehicle cannot be steered. If voltage at the powertrain control module is low or interrupted, however, the fuel shut-off may not occur and the vehicle can be accelerated while the steering system is locked. Also, when the control system shows that the ECL is unlocked and the vehicle is being driven, the lock pin location can vary. Depending on the location of the lock plate relative to the switch transition point, there could be contact between the lock plate and pin, causing the steering to lock while driving.
Consequence: If this occurs, a crash could result without warning.
Remedy: On vehicles equipped with an automatic transmission, the dealer will disable the steering column lock by removing the column-lock plate. When the ignition key is removed, the transmission shifter will lock but the steering column will not lock. On vehicles equipped with a manual transmission, the dealer will reprogram the powertrain control module, perform a dimensional check of the column lock, and, if necessary, replace the lock plate. The steering column on these vehicles will continue to lock when the key is removed. Owner notification to owners of 1997 vehicles equipped with automatic transmissions began on April 26, 2004. Owners of 1998–2004 vehicles (except 1997–1998 manual transmission vehicles) were notified beginning on August 2, 2004. The remaining owners (1997–1998 manual transmission vehicles) were notified beginning in late 2004. Contact Chevrolet at 1-800-630-2438.

1999
NHTSA campaign ID number: 00V111000
Recall date: April 6, 2000
Component: seatbelts: front
Potential number of units affected: 71,569
Summary: Vehicle description: passenger vehicles. The lap-belt webbing can twist, allowing the webbing to become jammed in the retractor.
Consequence: When the belt webbing becomes jammed in the retractor, the seatbelt may be unusable.
Remedy: Dealers will install inserts to the belt web guide of each lap-belt retractor.
Notes: Owner notification began November 13, 2000. Owners who take their vehicles to an authorized dealer on an agreed-upon service date and do not receive the free remedy within a reasonable time should contact Chevrolet at 1-800-222-1020. Also contact the National Highway Traffic Safety Administration's auto safety hotline at 1-888-327-4236.

NHTSA campaign ID number: 04V060000
Recall date: February 6, 2004
Component: Steering: column locking: antitheft device
Potential number of units affected: 126,624
Summary: On certain passenger vehicles equipped with electronic column-lock systems (ECL), when the ignition switch is turned to "lock," the ECL system prevents turning of the steering system. When the vehicle is started, the ECL unlocks the steering system. The vehicle is designed so that if the column fails to unlock when the vehicle is started and the customer tries to drive, the fuel supply will be shut off so that the vehicle cannot move when the vehicle cannot be steered. If voltage at the powertrain control module is low or interrupted, however, the fuel shut-off may not occur and the vehicle can be accelerated while the steering system is locked. Also, when the control system shows that the ECL is unlocked and the vehicle is being driven, the lock pin location can vary. Depending on the location of the lock plate relative to the switch transition point, there could be contact between the lock plate and pin, causing the steering to lock while driving.
Consequence: If this occurs, a crash could result without warning.
Remedy: On vehicles equipped with an automatic transmission, the dealer will disable the steering column lock by removing the column-lock plate. When the ignition key is removed, the transmission shifter will lock but the steering column will not lock. On vehicles equipped with a manual transmission, the dealer will reprogram the powertrain control module, perform a dimensional check of the column lock, and, if necessary, replace the lock plate. The steering column on these vehicles will continue to lock when the key is removed. Owner notification to owners of 1997 vehicles equipped with automatic transmissions began on April 26, 2004. Owners of 1998–2004 vehicles (except 1997–1998 manual transmission vehicles) were notified beginning on August 2, 2004. The remaining owners (1997–1998 manual transmission vehicles) were notified beginning in late 2004. Contact Chevrolet at 1-800-630-2438.

2000
NHTSA campaign ID number: 00V111000
Recall date: April 6, 2000
Component: seatbelts: front
Potential number of units affected: 71,569
Summary: Vehicle description: passenger vehicles. The lap-belt webbing can twist, allowing the webbing to become jammed in the retractor.
Consequence: When the belt webbing becomes jammed in the retractor, the seatbelt may be unusable.
Remedy: Dealers will install inserts to the belt web guide of each lap-belt retractor.
Notes: Owner notification began November 13, 2000. Owners who take their vehicles to an authorized dealer on an agreed-upon service date and do not receive the free remedy within a reasonable time should contact Chevrolet at 1-800-222-1020. Also contact the National Highway Traffic Safety Administration's auto safety hotline at 1-888-327-4236.

NHTSA campaign ID number: 04V060000
Recall date: February 6, 2004
Component: Steering: column locking: antitheft device
Potential number of units affected: 126,624
Summary: On certain passenger vehicles equipped with electronic column-lock systems (ECL), when the ignition switch is turned to "lock," the ECL system prevents turning of the steering system. When the vehicle is started, the ECL unlocks the steering system. The vehicle is designed so that if the column fails to unlock when the vehicle is started and the customer tries to drive, the fuel supply will be shut off so that the vehicle cannot move when the vehicle cannot be steered. If voltage at the powertrain control module is low or interrupted, however, the fuel shut-off may not occur and the vehicle can be accelerated while the steering system

is locked. Also, when the control system shows that the ECL is unlocked and the vehicle is being driven, the lock pin location can vary. Depending on the location of the lock plate relative to the switch transition point, there could be contact between the lock plate and pin, causing the steering to lock while driving.
Consequence: If this occurs, a crash could result without warning.
Remedy: On vehicles equipped with an automatic transmission, the dealer will disable the steering column lock by removing the column-lock plate. When the ignition key is removed, the transmission shifter will lock but the steering column will not lock. On vehicles equipped with a manual transmission, the dealer will reprogram the powertrain control module, perform a dimensional check of the column lock, and, if necessary, replace the lock plate. The steering column on these vehicles will continue to lock when the key is removed. Owner notification to owners of 1997 vehicles equipped with automatic transmissions began on April 26, 2004. Owners of 1998–2004 vehicles (except 1997–1998 manual transmission vehicles) were notified beginning on August 2, 2004. The remaining owners (1997–1998 manual transmission vehicles) were notified beginning in late 2004. Contact Chevrolet at 1-800-630-2438.

2001
NHTSA campaign ID number: 04V060000
Recall date: February 6, 2004
Component: Steering: column locking: antitheft device
Potential number of units affected: 126,624
Summary: On certain passenger vehicles equipped with electronic column-lock systems (ECL), when the ignition switch is turned to "lock," the ECL system prevents turning of the steering system. When the vehicle is started, the ECL unlocks the steering system. The vehicle is designed so that if the column fails to unlock when the vehicle is started and the customer tries to drive, the fuel supply will be shut off so that the vehicle cannot move when the vehicle cannot be steered. If voltage at the powertrain control module is low or interrupted, however, the fuel shut-off may not occur and the vehicle can be accelerated while the steering system is locked. Also, when the control system shows that the ECL is unlocked and the vehicle is being driven, the lock pin location can vary. Depending on the location of the lock plate relative to the switch transition point, there could be contact between the lock plate and pin, causing the steering to lock while driving.
Consequence: If this occurs, a crash could result without warning.
Remedy: On vehicles equipped with an automatic transmission, the dealer will disable the steering column lock by removing the column-lock plate. When the ignition key is removed, the transmission shifter will lock but the steering column will not lock. On vehicles equipped with a manual transmission, the dealer will reprogram the powertrain control module, perform a dimensional check of the column lock, and, if necessary, replace the lock plate. The steering column on these vehicles will continue to lock when the key is removed. Owner notification to owners of 1997 vehicles equipped with automatic transmissions began on April 26, 2004. Owners of 1998–2004 vehicles (except 1997–1998 manual transmission vehicles) were notified beginning on August 2, 2004. The remaining owners (1997–1998 manual transmission vehicles) were notified beginning in late 2004. Contact Chevrolet at 1-800-630-2438.

Notes
GM recall number 04006: Customers can also contact the National Highway Traffic Safety Administration's auto safety hotline at 1-888-327-4236.

2002
NHTSA campaign ID number: 04V060000
Recall date: February 6, 2004
Component: Steering: column locking: antitheft device
Potential number of units affected: 126,624
Summary: On certain passenger vehicles equipped with electronic column-lock systems (ECL), when the ignition switch is turned to "lock," the ECL system prevents turning of the steering system. When the vehicle is started, the ECL unlocks the steering system. The vehicle is designed so that if the column fails to unlock when the vehicle is started and the customer tries to drive, the fuel supply will be shut off so that the vehicle cannot move when the vehicle cannot be steered. If voltage at the powertrain control module is low or interrupted, however, the fuel shut-off may not occur and the vehicle can be accelerated while the steering system is locked. Also, when the control system shows that the ECL is unlocked and the vehicle is being driven, the lock pin location can vary. Depending on the location of the lock plate relative to the switch transition point, there could be contact between the lock plate and pin, causing the steering to lock while driving.
Consequence: If this occurs, a crash could result without warning.
Remedy: On vehicles equipped with an automatic transmission, the dealer will disable the steering column lock by removing the column-lock plate. When the ignition key is removed, the transmission shifter will lock but the steering column will not lock. On vehicles equipped with a manual transmission, the dealer will reprogram the powertrain control module, perform a dimensional check of the column lock, and, if necessary, replace the lock plate. The steering column on these vehicles will continue to lock when the key is removed. Owner notification to owners of 1997 vehicles equipped with automatic transmissions began on April 26, 2004. Owners of 1998–2004 vehicles (except 1997–1998 manual transmission vehicles) were notified beginning on August 2, 2004. The remaining owners (1997–1998 manual transmission vehicles) were notified beginning in late 2004. Contact Chevrolet at 1-800-630-2438.

2003
NHTSA campaign ID number: 04V060000
Recall date: February 6, 2004
NHTSA campaign ID number: 04V060000
Recall date: February 6, 2004
Component: Steering: column locking: antitheft device
Potential number of units affected: 126,624
Summary: On certain passenger vehicles equipped with electronic column-lock systems (ECL), when the ignition switch is turned to "lock," the ECL system prevents turning of the steering system. When the vehicle is started, the ECL unlocks the steering system. The vehicle is designed so that if the column fails to unlock when the vehicle is started and the customer tries to drive, the fuel supply will be shut off so that the vehicle cannot move when the vehicle cannot be steered. If voltage at the powertrain control module is low or interrupted, however, the fuel shut-off may not occur and the vehicle can be accelerated while the steering system

is locked. Also, when the control system shows that the ECL is unlocked and the vehicle is being driven, the lock pin location can vary. Depending on the location of the lock plate relative to the switch transition point, there could be contact between the lock plate and pin, causing the steering to lock while driving.
Consequence: If this occurs, a crash could result without warning.
Remedy: On vehicles equipped with an automatic transmission, the dealer will disable the steering column lock by removing the column-lock plate. When the ignition key is removed, the transmission shifter will lock but the steering column will not lock. On vehicles equipped with a manual transmission, the dealer will reprogram the powertrain control module, perform a dimensional check of the column lock, and, if necessary, replace the lock plate. The steering column on these vehicles will continue to lock when the key is removed. Owner notification to owners of 1997 vehicles equipped with automatic transmissions began on April 26, 2004. Owners of 1998–2004 vehicles (except 1997–1998 manual transmission vehicles) were notified beginning on August 2, 2004. The remaining owners (1997–1998 manual transmission vehicles) were notified beginning in late 2004. Contact Chevrolet at 1-800-630-2438.

2004

NHTSA campaign ID number: 04V273000
Recall date: June 3, 2004
Component: steering: linkages
Potential number of units affected: 41,928
Summary: On certain passenger vehicles involved in this campaign, the lower control arm ball-stud nut/washer assemblies with washers were made of the wrong material. The washers may fracture and become loose or fall away from the vehicle, reducing clamp load. Separation of the control-arm ball stud and steering knuckle, due to disengagement of the tapered attachment and retaining nut, is possible.
Consequence: If the control arm separates from the knuckle, the affected corner of the vehicle will drop and the control arm will be forced downward, contacting the wheel. The affected wheel could tilt outward and create a dragging action that would slow the vehicle and create a tendency for the vehicle to turn in the direction of the affected wheel. In extreme situations, the affected wheel assembly could separate from the vehicle. Separation of the wheel assembly would also sever that wheel's hydraulic brake hose and result in diminished braking performance of the vehicle, which could result in a crash.

Remedy: Dealers will inspect the ball-stud joints and measure the torque of the nut for all front lower control arms in all models affected, as well as the rear lower control arms on the Corvette. If the torque is not at specification, the dealer will replace the ball stud, the knuckle, and the nut. If the torque is at specification, the dealer will replace the nut/washer assembly only. This recall began on August 25, 2004. Owners should contact Chevrolet at 1-800-630-2438.

NHTSA campaign ID number: 04V060000
Recall date: February 6, 2004
Component: Steering: column locking: antitheft device
Potential number of units affected: 126,624
Summary: On certain passenger vehicles equipped with electronic column-lock systems (ECL), when the ignition switch is turned to "lock," the ECL system prevents turning of the steering system. When the vehicle is started, the ECL unlocks the steering system. The vehicle is designed so that if the column fails to unlock when the vehicle is started and the customer tries to drive, the fuel supply will be shut off so that the vehicle cannot move when the vehicle cannot be steered. If voltage at the powertrain control module is low or interrupted, however, the fuel shut-off may not occur and the vehicle can be accelerated while the steering system is locked. Also, when the control system shows that the ECL is unlocked and the vehicle is being driven, the lock pin location can vary. Depending on the location of the lock plate relative to the switch transition point, there could be contact between the lock plate and pin, causing the steering to lock while driving.
Consequence: If this occurs, a crash could result without warning.
Remedy: On vehicles equipped with an automatic transmission, the dealer will disable the steering column lock by removing the column-lock plate. When the ignition key is removed, the transmission shifter will lock but the steering column will not lock. On vehicles equipped with a manual transmission, the dealer will reprogram the powertrain control module, perform a dimensional check of the column lock, and, if necessary, replace the lock plate. The steering column on these vehicles will continue to lock when the key is removed. Owner notification to owners of 1997 vehicles equipped with automatic transmissions began on April 26, 2004. Owners of 1998–2004 vehicles (except 1997–1998 manual transmission vehicles) were notified beginning on August 2, 2004. The remaining owners (1997–1998 manual transmission vehicles) were notified beginning in late 2004. Contact Chevrolet at 1-800-630-2438.